Seed Production of Agricultural Crops

Other Related Books

Cowpea in India	*Kumar, D.*
Crops of India	*Das, N.R.*
Genetic Improvement of Field Crops	*Singh, C.B.*
Guar In India	*Kumar, D.*
Hand Book of Seed Industry (Prospects & its costing)	*Ramdeo, A.*
Handbook of Seed Science and Technology	*Basra, A.S.*
Mothbean in India	*Kumar, D.*
Organic Seed: Indigenous Varieties and Technologies	*Vanangamudi, K.*
Principles and Practices of Seed Storage	*Justice, O.L.*
Seed Borne Diseases - Ecofriendly Management	*Arya, Arun*
Seed Borne Diseases Objectionable in Seed Production and their Management	*Bhale, M.S.*
Seed Economics: Commercial Considerations for Enterprise Management in Developing Countries	*Kugbei, S. (ICARDA)*
Seed Identification Manual	*Martin, A.C.*
Seed Pathology (Vol. 1-2 set)	*Neergaard, P.*
Seed Quality Enhancement Principles and Practices	*Vanangamudi, K.*
Seed Science and Technology: Laboratory Manual	*McDonald, M.B.*
Seed Technology	*Khare, D.*

Seed Production of Agricultural Crops

A. Fenwick Kelly

SCIENTIFIC
PUBLISHERS (INDIA)

Reprinted in India, 2013
by
Scientific Publishers (India)

5-A, New Pali Road,	4806/24, Ansari Road,
P.O. Box 91	Daryaganj
Jodhpur - 342 001 (India)	New Delhi - 110 002 (India)

www.scientificpub.com

Print: 2020

ISBN: 978-81-7233-818-3

First Published 1988

Printed in India

To Mickie

CONTENTS

FOREWORD

The potential for improved new crop varieties can only be achieved in practice when farmers, skilled in producing seed crops, can convert the breeders' genetic material into marketable seed-lots for crop production.

The author of *Seed Production of Agricultural Crops*, A. Fenwick Kelly, has unique knowledge and experience following his career at NIAB where his work has considerably influenced the present methods of variety identification, variety trialling and all aspects of seed production.

This book has the attribute of combining the principles of seed production with all the practical requirements needed for producing quality seed for a range of arable and fodder crops in the field. It fills a much-needed gap, bringing together all the botanical, seed-growing, quality-control and marketing aspects in producing farm seeds. All these aspects are essential components to ensure that crop breeders' efforts can be translated into improved crop performance for all farmers, whether in the temperate or tropical regions.

Graham Milbourn, MSC PhD,
Director, National Institute of Agricultural Botany, Cambridge
February 1987

PREFACE

Seed is fundamental to the production of crops. Much has been written about the importance of using seed which is viable and free from weed seeds and diseases. However, rather little has been said about the need to preserve the genetic quality of seed during multiplication. Yet with the production of ever more sophisticated cultivars the work of the seed producer becomes of increasing importance. High seed yields are not enough in themselves and we need to arrange production in such a way that the qualities which give each cultivar a special place in agricultural production are preserved from one generation to the next. This book therefore lays emphasis on the precautions which the seed producer should take to ensure the production of good-quality seed in the widest sense.

During my work at the National Institute of Agricultural Botany (NIAB) in Cambridge I was privileged to have the opportunity to study seed production both as a research subject and as a practical exercise. I also travelled widely in Europe and North America and am greatly indebted to the many people who discussed with me the various problems, beginning with the definition of a cultivar and ending with the stored seed ready for planting. The book is essentially practical, but contains enough basic information to enable the reader to understand the reasons behind the management practices which are described. It is hoped that it will be of value to those in agricultural education who extend crop production into seed production, to extension workers and administrators in seed programmes and to the practical seed producers.

I am grateful for the help and encouragement of Dr Graham Milbourn, the Director and the staff of the NIAB who have filled in many of the details. Likewise, I acknowledge the help received from Dr Walter P. Feistritzer, and the staff of the Seed Service, Plant Production and Protection Division of the UN Food and Agriculture Organisation. Finally, it is a great pleasure to acknowledge the encouragement of Raymond George of Bath University, who first proposed that I should write the book and has given me every encouragement throughout.

A. Fenwick Kelly
Cambridge, October 1986

ACKNOWLEDGEMENTS

We are indebted to the following for permission to reproduce copyright material:

Food and Agriculture Organisation of the United Nations for tables 5.1 and 5.2 from table 1, p. 41 and table 7, p. 98; International Board for Plant Genetic Resources for descriptors (IBPGR, 1980–1985); National Institute of Agricultural Botany and ISTA for table 3.3 (ISTA, 1985) and NIAB Library for formula on p. 159 (NIAB, 1982a) and table on p. 201 from p. 2 (NIAB, 1975); Organisation for Economic Co-operation and Development, Paris for figs. 1.1 (Cowling, Kelly and Demarly, 1960) and 2.6 (OECD, 1982); Shelbourne Reynolds Engineering Ltd. for figs. 3.1 and 3.2; Shell Chemicals U.K. Ltd. for fig. 3.5; Toveys of Cirencester Ltd. for table 3.2 and fig. 3.3; Vogelenzang Andelst B.V., The Netherlands for fig. 5.1; the author, John T. Ward for table 9.1 from table 14.2 (Ward *et al.*, 1985).

PART 1 PRINCIPLES OF SEED PRODUCTION

1 REPRODUCTION OF PLANTS

SEED

In nature, seeds overcome three major problems for the plant. First, they provide the method by which a plant can multiply, because a single plant is capable of producing many, often very many, seeds. Second, they provide a means by which the plant can survive adverse conditions, often for long periods, because seeds are able to survive in the soil and each seed is provided with mechanisms which enable it to be dormant until the right conditions for germination occur. And third, seeds provide the plants with the possibility to move in space because they may be dispersed after ripening by various means such as wind, water or contact with animals or birds.

In agriculture, the art of seed production lies in harnessing these properties to provide for human needs. The aim is to multiply desirable seeds to provide for future crops. This is done by establishing a seed crop with a sufficient number of the desired plants to produce the required quantity of seed at the estimated rate of multiplication. Subsequently the seed crop is harvested at the right time and the seed is cleaned and stored. From store it can be dispersed for sowing new crops in a following season.

To exploit these natural assets to the full we need to understand some of the processes by which a plant sets seed and by which that seed remains viable until required. In this way, seed-crop management can be adjusted to secure the best possible results.

REPRODUCTIVE SYSTEMS

There are two main ways in which plants are reproduced for use in agriculture. Some crop species are reproduced vegetatively: parts of stems or roots may develop into specialised organs – for example the potato tuber – which can be used in a similar manner to seed; alternatively some crops are propagated by stem cuttings – for example, cassava – which are encouraged to develop roots. Vegetative propagation has the advantage that plants propagated in this way are more stable genetically than those reproduced through seed, and techniques have now been developed which reduce to a minimum the risk of spreading disease via the vegetative plant parts used for propagation. The main disadvantage of the system is that the vegetative plant parts are more difficult to handle, store and transport than seed, and they are not as long lasting; it is rarely possible to store vegetative plant parts beyond the next growing season whereas seed can be retained for longer periods.

The second system of reproduction is by a sexual process through seed. Sexual reproduction may be achieved either by self-fertilisation or cross-fertilisation. A few crop species reproduce by a process known as 'apomixis' which gives seed-like bodies which behave in all respects like normal seed.

When apoximis occurs the seed is produced without the process of fertilisation. This may be by agamospermy, where the egg-cell develops, or by vivipary, where vegetative proliferations are produced. In the agricultural crops apomixis is generally confined to some of the grass species – for example *Poa pratensis* L. among the temperate grasses, or *Cenchrus cilaris* L. among the tropical species. For some species apomixis is obligatory, but in others it may occur only to some extent, with sexual reproduction being more or less important. In practice, apomixis has similar implications for seed production to self-fertilisation.

Self-fertilisation occurs when pollen from the anthers of a flower is transferred to the stigma of that same flower; when this pollen germinates and comes in contact with the embryo-sac, self-fertilisation is complete. In some crop species, self-fertilisation can occur before the flowers open when the flowers are said to be 'cleistogamous'. In these circumstances the degree of self-fertilisation is absolute. However, in many crop species which are normally self-fertilised the flowers do open before pollination, and even in those species in which cleistogamy occurs there can be certain effects which cause the flowers to open. When this happens, some degree of cross-fertilisation is possible. Many of the winter barley cultivars grown in northern Europe are prone to some degree of cross-fertilisation although barley is regarded as usually a self-fertilising species.

In seed growing, the main advantage of self-fertilisation is the greater degree of genetic stability which it confers. Self-fertilised species are more likely to produce seeds which are true to the cultivar characteristics than those which are cross-fertilised because the genetic background of male and female is the same.

Cross-fertilisation occurs when pollen from one flower is transferred to the stigma of another flower to effect fertilisation. The flower in which fertilisation takes place may be on the same plant as the one from which the pollen originates, or on a different plant. For cross-pollination to occur there must be some way in which pollen can be transported from one flower to another, and to ensure cross-fertilisation there needs to be some mechanism which mitigates against self-fertilisation. In the agricultural crop species pollen is transported either by wind or by insects. Those species which rely upon wind usually have erect flowering stems so that the wind is able easily to disperse the pollen and the receptive stigmas are not too encumbered by leaves. Species which rely upon insects, on the other hand, have flowers which are more elaborate and which attract insects with scents, coloured petals and a good supply of nectar.

For the seed grower the method by which cross-pollination is achieved is important. If wind is the transporting agent, consideration needs to be given to the direction and velocity of the prevailing winds in the area where the seed crop is to be grown, and account should be taken of the physical features of the landscape which might affect the wind – for example groups of tall trees or high hedges. If insects are the main agents for cross-pollination it is essential to know which kind of insects are the most efficient pollinators, and to be sure that the area in which the seed crop is to be grown is a suitable habitat for them. For example, humble-bees are one of the main pollinating agents for *Trifolium pratense* L., and to ensure a good seed crop the surrounding area should provide enough winter nesting sites for the queens in rough ground, otherwise the humble-bee population may be inadequate to pollinate the crop. For some crops it may be possible to introduce colonies of pollinating insects – for example honey-bees are often used for lucerne seed crops.

Mechanisms are needed to mitigate against self-fertilisation because of the vast amounts of pollen produced by each flower. Species in which these mechanisms occur are said to be 'self-incompatible'. In the wind-pollinated species, self-incompatibility is usually achieved by allowing the anthers to develop and shed pollen before the stigmas are receptive. An extreme example of this occurs in maize where the male flower is separate from the female, and occurs at the top of the plant: the male tassel usually matures in advance of the female silk so that the chance is improved that fertilisation will be effected by pollen from a different plant.

When insects are the pollinating agents, self-incompatibility is often achieved through the structure of the flowers. The insect is guided by the position of the nectaries or other means in such a way that pollen from the anthers is scattered over its body in places which render it most likely that the pollen will be deposited on a stigma of the next flower visited.

Apart from such mechanical devices, self-incompatibility can also be achieved through genetic control. Pollen from a particular flower is unable after germination to penetrate the stigma of that flower, or may be inhibited from growing within the style; nevertheless it is not inhibited from effecting pollination if transferred to a different flower. Self-incompatibility is not necessarily always complete. Just as in normally self-fertilising species there can be some degree of cross-fertilisation, so self-fertilisation is often possible in normally cross-fertilising species.

The maintenance of trueness-to-cultivar is more difficult in cross-fertilising crops than in those which are self-fertilised. The male parent will normally have a different genetic background from the female, and the source of the pollen which actually effects fertilisation is less certain since it must be transported over some distance to reach the female. There is thus greater opportunity for pollen from an unknown source to effect fertilisation. For the seed grower, these are important considerations in the siting of seed crops so as to give the greatest possible chance that pollination and subsequent fertilisation will in fact occur with pollen from the desired male parent.

In both methods of sexual reproduction – self- or cross-fertilisation – there is a possibility that mutation will occur. Some mutants will be easily observable in the seed crop as, for example, are speltoids in wheat. Others may not be seen in the outward appearance of the plant. The possibility of mutation thus has to be considered in relation to standards for trueness-to-cultivar which can be achieved.

ENVIRONMENTAL CONDITIONS

The growth of a seed crop can be affected in various ways by the environment. The length of the growing season must be sufficient for the crop to mature; that is to say that the time from sowing to harvesting must be sufficient for the crop to flower, set seed and ripen that seed. The start of the growing season – sowing time – will be governed by temperature and moisture supply; the soil must be warm enough and moist enough to permit the sown seed to germinate. Seeds of crops adapted to temperate climates will germinate at lower temperatures than those of crops adapted to the tropics. For example, the use of maize as a grain crop in northern Europe is restricted by the inability of the seed to germinate satisfactorily at the relatively low soil temperatures prevalent in early spring so that it must be sown later when the soil has warmed; this results in a short growing season giving insufficient time for the crop to ripen except in the most favoured areas.

During the growing season, temperature and moisture must remain suitable to sustain growth, and maturity of the seed crop must occur when rainfall is not excessive so that seed can be harvested when it is reasonably dry. It is also necessary that within the growing period suitable conditions will occur to stimulate the plants to flower and set seed. This change from a vegetative phase, when the plant produces mainly leaves above ground and roots below, to a reproductive phase, when it produces flowering stems, flowers and eventually seeds is a response to external stimuli.

Most species will not change to the reproductive phase until they have completed some vegetative growth and have reached a stage called 'ripeness to flower'. Thereafter there are two main stimuli which cause the change to occur: temperature and day-length.

The effect of temperature is twofold. Obviously the temperature at the time that shoot growth starts must be high enough or growth will be inhibited. However, for many crops, particularly those which are perennial or biennial (including those normally sown in the autumn) it is the temperature during the period before the change occurs that is important: they require a period of low temperature to stimulate the reproductive phase. This is known as 'vernalisation'. The length of the cold spell needed varies between crops and between cultivars within crops. For example, some wheat cultivars can be sown later than others, and will still flower and set seed. However, the autumn-sown cultivars always require some degree of vernalisation whereas the true spring-wheat cultivars do not. The seed grower has to adjust the time of sowing a seed crop so as to ensure that the plants reach ripeness to flower at a time which will still allow them to be vernalised before the reproductive phase commences.

The second main agent causing plants to flower is day-length, or the length of the dark period each 24 hours. Plant species are termed 'short day', 'long day' or 'day neutral'. Short-day plants are those adapted to flower when there is little variation between the length of daylight and the length of darkness. Long-day plants require a length of daylight greater than the length of darkness before they can flower. Day-neutral plants are those which are indifferent to the length of day and will flower at any latitude. A short-day plant can be inhibited from flowering at a higher latitude, and similarly a long-day plant will not flower if moved nearer to the tropics. The main concern for the seed grower is to ensure that the crop it is intended to grow will be capable of flowering under the day-lengths where the crop is to be grown. The influence of day-length on time of flowering in *Lolium perenne* L. is illustrated in Fig. 1.1 which shows the effect of moving cultivars from higher to lower latitude. The range of heading dates was much greater at the lower latitudes because the earlier cultivars, i.e. those requiring less stimulus from longer days, headed much earlier. The later cultivars were delayed and it was noted that some plants of these cultivars failed to head at the lower latitudes.

Thus there can be a complex interaction of various factors which cause plants to change from the vegetative to the reproductive phase and subsequently to extend flowering stems. The plants must be of the correct age, must have been subjected to certain temperature changes and have received the required stimulus in respect to photo-period. The seed grower has to adjust management to ensure that these conditions are met for the particular crop species being grown. Once flower initiation has begun the seed crop should continue to grow vigorously. The supply of nutrients and moisture should be adequate. However, there is evidence that in many species excess

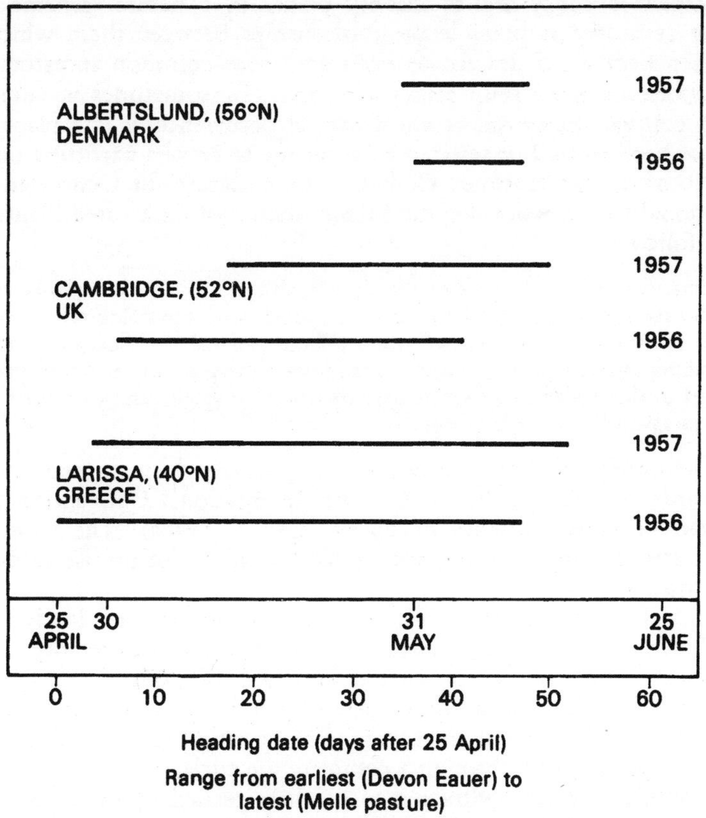

Fig. 1.1 Effect of latitude on heading date of cultivars of perennial ryegrass. Range from earliest (Devon Eaver) to latest (Melle Pasture). (Based on data in Cowling, Kelly and Demarly, 1960)

nitrogen at this time can delay and reduce flowering, whereas shortage of nitrogen may cause earlier flowering.

When flowers have developed, the next stage is to ensure so far as possible that conditions are satisfactory for pollination. Except for cleistogamous species a crop which is too well grown and lodged at the time of pollen release will not be as satisfactory as a standing crop. Plant-to-plant spacing is important in some crops and the spacing which best suits the growing of a food crop is not necessarily the most suitable for a seed crop.

Weather at the time of pollination is important. Generally pollen does not survive well in heavy rain, and extremes of temperature can cause poor pollination. Insects, particularly bees, will not work well in very wet weather. Sites for seed crops should thus be chosen which will be likely to provide suitable conditions at pollination time.

THE CONCEPT OF CULTIVAR

The botanist is concerned to classify plants in a natural scheme which expresses as nearly as possible the relationships between them which may have arisen because of descent or evolution from common ancestors. Such classifications usually group plants in species, or sometimes in subspecific units. A cultivar, however, is not a natural occurrence; it is a plant group which has been created or selected by someone to serve a particular purpose. Article 10 of the International Code of Nomenclature for Cultivated Plants (International Commission for the Nomenclature of Cultivated Plants 1980) reads as follows:

> The international term *cultivar* denotes an assemblage of cultivated plants which is clearly distinguished by any characters (morphological, physiological, cytological, chemical, or others) and which, when reproduced (sexually or asexually), retains its distinguishing characters. The cultivar is the lowest category under which names are recognised in this code. The term is derived from *culti*vated *var*iety or their etymological equivalents in other languages.

This definition refers to a cultivar as being 'clearly distinguished' and these same words are used in Article 6 of the International Convention for the Protection of New Varieties of Plants (UPOV 1978a). The Convention further states that the characteristics 'must be capable of precise recognition and description'.

As might be expected, therefore, these two definitions are closely similar, but both require subjective assessment of the margin of difference between two cultivars which can be regarded as 'clearly distinguished'.

A cultivar is defined as an assemblage of plants, but in fact the characters which distinguish it can only be observed by sowing the seed and measuring them as the plants grow through a complete life cycle.

For example, a wheat cultivar may be characterised by the colour of the seed, the growth habit of the seedlings, the plants' resistance to certain diseases and the shape of the ears. This is a picture built up over a period of time, but it cannot be observed as a whole at one and the same time.

Furthermore, the original unit on which the observations were made will have been destroyed: the seed has been used to grow plants, and at the end a new generation of seed has been created. This argues that the differences between cultivars cannot be very small because there is likely to be some variation in the expression of the distinguishing characteristics between one plant and another and between one generation and another, especially in those cultivars which are cross-fertilised.

Although the *International Code* makes no reference to homogeneity, the *Convention* states that a cultivar must be 'sufficiently homogeneous, having regard to the particular features of its sexual reproduction or vegetative propogation'. Thus, while in the nature of the material it is inevitable that the plants constituting a cultivar will show some variability in the characteristics which distinguish that cultivar from others, some degree of homogeneity is required otherwise it becomes impossible to distinguish the material as a single cultivar. Homogeneity also has a bearing on stability. It becomes much more difficult to retain a cultivar's distinguishing characteristics when it is reproduced if the material is not sufficiently homogeneous.

Thus the concept of what material validly constitutes a cultivar depends mainly upon the breeding system of the species concerned and upon the methods employed by the plant breeders. When growing seed, we need to know what kind of material is to be multiplied as this may affect the stability of the material from one generation to the next and consequently its usefulness in crop production.

The landrace or local cultivar

Before the advent of plant breeding as a means of creating new cultivars, two forces were at work which influenced the types of crop plants available. First was the effect of natural selection imposed by the environment: climatic effects caused by changes in day-length, temperature and rainfall; soil effects caused by nutritional status, presence of toxic substances, etc.; and effects of predators or pathogens naturally occurring in the vicinity. Second was the effect of the conscious or unconscious selection by the person growing the crop. Conscious selection might be exercised through the deliberate selection of better plants or plumper seeds to provide the seed for the next crop. Unconscious selection would result from the management practices imposed upon the seed crop; for example, time of sowing or harvesting could influence the proportion of seed contributed by certain individual plants within the population. These influences have combined to create a wealth of locally adapted cultivars, many of which have highly valuable characteristics. Having evolved over a period of years, these local cultivars are usually very stable so long as they are maintained within the area of origin and are subjected to traditional management practices; stability of performance rather than spectacular yield is their main asset. However, the need to achieve higher yields to provide food for an increasing population has generally made it necessary artificially to create variability within which selections can be made leading to new, improved cultivars with higher yield potential and other desirable characteristics.

Plant-breeding methods leading to different types of cultivars

In normally self-fertilising crops the usual method of creating new variability within which a cultivar may be selected is by hybridisation. By this method the desirable characteristics of two or more parents may be combined. The type of cultivar obtained will depend in part upon the genetic background of the chosen parents, and in part upon the method of selection employed after the final cross has been made. A similar situation arises when induced mutations are used to create new variability.

Classic plant breeding was based on the 'pure line' theory of Professor W. Johannsen (Copenhagen 1903). This theory defined a pure line as all the descendants of a single homozygous individual by continued self-fertilisation. The result would be a homogeneous cultivar. However, after hybridisation there is considerable heterogeneity in the resultant progeny, and to multiply this in bulk in order to be able to select homozygous individuals would be a very lengthy process. The selection of most modern cultivars therefore begins at an early stage – F2 – and they are subsequently selected as lines. They are probably released at about F8 to F12, and are therefore not as homogeneous as a pure line would be.

An extension of the line method of selection is to produce multiline cultivars. Normal line selection aims to create a new cultivar on the basis of one line or a few lines which are closely similar. In the multiline cultivar one or more of the lines may be different from the others in an essential characteristic – usually in the resistances to particular diseases which have been incorporated. By incorporating different sources of resistance the cultivar is buffered against changes in the virulences of the pathogen. However, such cultivars may be less stable than those selected by conventional methods because a change in the prevalence of the virulences of the pathogen may eliminate certain lines from the cultivar, so necessitating a return to the plant breeder so that the cultivar can be reconstituted. Some plant breeders claim that this is an advantage because it enables them to substitute new resistance sources in their material when it occurs; however, this raises the question as to whether the new version of the cultivar can be regarded as the same or a new and different one.

An alternative plant-breeding method is to create a composite cross by bulking the F2 generation of several crosses. The composite may be allowed to develop for several generations during which natural selection may occur. If the composite is grown at several locations, in time locally adapted selections will be developed. The composite also forms a gene pool and selections can be made by the plant breeder from within the composite after several generations of multiplication.

An alternative to the composite is the synthetic, in which a number of lines are put together by the plant breeder in predetermined proportions. Generally, the synthetic has a limited life since the proportions of the constituent lines are likely to change over a number of generations. Seed production thus has to be planned on a limited-generation basis. An extension of this system is to use mixtures of cultivars; advantages have been claimed in some species for a mixture of cultivars over a single cultivar, particularly if different resistance genes are present in each cultivar. However, the costs of mixing need to be taken into account, and to cheapen mixtures it has been suggested that they might be grown for seed for one or two generations after mixing before being used for crop production, but the advantages of this procedure are at present less clear.

A hybrid cultivar is the result of a controlled cross between a male and a female parent; the seed is harvested only from the female parent and is used for crop production. In the self-fertilising species the production of hybrid cultivars is possible if male sterile lines can be found for use as female parents. Recent developments in the agro-chemical industry have produced materials which can be applied to wheat so as to prevent self-pollination; this may be a gametocide which destroys the pollen of the desired female parent, or an inhibitor which prevents pollen produced by the female parent from effecting fertilisation. The advantage of the hybrid cultivar lies in heterosis. Special measures are required to produce seed, which can be expensive; seed is harvested from a proportion of the seed crop only – the female parent. Heterosis, therefore, has to produce a considerable effect to make the production of hybrid cultivars in self-pollinating species worth while.

In cross-pollinating species, the plant breeder usually looks for parent plants which have good combining ability, that is, plants which when allowed to multiply together produce a desirable combination of characteristics. By the nature of the material, heterozygosity will be greater in cross-fertilised species than in those which are self-fertilised, and they will in general be less homogeneous; each generation of an open-pollinated cultivar is in effect a mixture of hybrids. Usually, therefore, open-pollinated cultivars are grown on a limited-generation basis and it is necessary to return to the plant breeder's maintenance material after each cycle of seed production to produce commercial quantities.

Composites can be created by putting together a large number of parent plants and allowing random pollination to occur. Generally a composite in a cross-fertilising species should be the first generation of such a random pollination.

A synthetic starts with a limited number of specified parents which are then permitted to inter-pollinate. The number of generations of multiplication is strictly limited so as to re-create the cultivar at the end of each multiplication cycle.

As with self-fertilised species, hybrid cultivars of cross-fertilised species are created by controlling pollination to ensure that seed is produced from a desired crossing. This can be achieved in three different ways.

1. The female parent can be emasculated. This can be achieved in monoecious plants, for example maize, by removing the male flowers before pollen is released.
2. By using male sterility in the female line the same result can be achieved without the need physically to remove the male flowers.
3. Another possibility is to use self-incompatibility. In some crops using this system the seed crop is harvested as a whole since all plants are contributing and receiving pollen. This method is used in some brassicas. The main disadvantage is that self-incompatibility is not always complete so that some inbred plants may be produced. Where these are excessive the heterosis advantage in the subsequent crop is diminished.

The advantage of the hybrid cultivar in cross-pollinated species does not lie only in heterosis. Most hybrids are based upon inbred lines. Several generations of inbreeding of a normally cross-fertilising plant will reduce heterozygosity and make possible the inclusion of desirable genes; the controlled cross between two such inbreds produces heterosis, and a desirable combination of genes.

The disadvantage is the higher cost of plant breeding and of seed production. Considerable work is required to produce desirable inbreds which can only be used for making new hybrid cultivars; they have no value for producing a food crop. The final seed crop is not fully productive when male sterility or emasculation are used because only the female parent is harvested for seed.

For this reason, various other hybrids have been produced. The hybrid resulting from the use of two inbred lines is a 'single cross'. The 'double cross' hybrid is the F1 resulting from the use of two single cross hybrids as parents. In a 'three-way cross', an inbred is mated with an F1 hybrid. A 'top cross' is the F1 resulting from a cross between an inbred or a single cross and an open-pollinated cultivar. These various forms of hybrid cultivar each require a particular cycle of seed production to produce the seed used in crop production.

2 SEED QUALITY AND QUALITY CONTROL

ASPECTS OF QUALITY

The quality of the seed which the farmer uses is all important in deciding whether a crop will be good, bad or indifferent. There are, however, several different aspects of quality, each of which can affect a crop.

The most important is 'genetic quality' because this is what ensures that the plants making up the crop possess the desired characteristics. Other aspects of quality which can affect the growth of the crop are: quality of viability and germination; analytical quality; health quality; and physical quality. In addition, seed needs to have good storage quality to ensure that it maintains condition until it is used for sowing.

The genetic quality is generally controlled in the first instance by cultivar trials which are described in outline in the following section; for those wishing to study the techniques further, see Bibliography and References. Seed production of selected cultivars is controlled by plot tests on samples of the seed and by field inspection. The descriptions of the techniques for these procedures are based on a paper prepared by the author for the Organisation for Economic Co-operation and Development (OECD) and subsequently published as a pamphlet entitled Methods for Plot Tests and Field Inspection (OECD 1982) in connection with the OECD seed schemes. References have been added for those interested in the statistical techniques used in deriving the various quality-control sampling procedures.

Quality attributes such as germination capacity, purity, health, and moisture content are determined on samples of seed in the laboratory. Seed-testing rules have been published by the International Seed Testing Association (ISTA) in 1985. The ISTA has as its primary purpose the development, adoption and publishing of standard procedures for sampling and testing seeds and the promotion of uniform application of those procedures. The ISTA rules are widely used in seed-testing laboratories throughout the world, and the reader is referred to these (see Bibliography and References) for the details of seed sampling and testing procedures.

GENETIC QUALITY

As we saw in the previous chapter, the aim of the seed producer is to multiply desirable seeds to provide for future crops. Perhaps the most important aspect of quality or desirability lies in the genetic background of the seed: what kind of plant will it produce within the next crop? The kind of plant, and its characteristics, are programmed into the seed at the time of fertilisation. For ease of handling, the description of the plants which a seed-lot will produce is condensed into the cultivar name. A person who buys seed of a named cultivar expects that the crop grown from such seed will show particular characteristics of plant form which, when harvested, will give produce of a particular quality. Unless the seed has the potential to perform in this way no effort of good husbandry will be able to produce a satisfactory crop.

The first step in safeguarding genetic quality, therefore, is to identify cultivars which show desirable characteristics. There are two aspects to this: first we need to determine the value of the cultivar for cultivation and for the usefulness of the harvested product; second, we need to determine some relatively quick and easy way to identify seed-lots or seed crops during multiplication, which may extend over several generations.

Cultivar evaluation for cultivation and use

To identify cultivars which show desirable characteristics requires some systematic testing procedure. Potentially useful cultivars have to be tested in field trials designed to identify those which have a combination of characteristics most likely to give enhanced value for cultivation and use. Individual plant breeders begin this process at an early stage and progressively eliminate the less desirable material: a single cross of a self-fertilised crop species may produce several thousand potential cultivars, from among which only one or two showing the desired combination of characteristics have to be selected; similarly, selection of desirable parents within a cross-fertilised crop species involves testing very many possible combinations. At a certain point the most promising material is included in trials, where comparison can be made with cultivars which are already in use and with material from other plant breeders working within an area with similar agro-ecological conditions. Such areas are not necessarily confined within national boundaries and it is not unusual to find cultivars which can be grown in several countries; for example, many cereal cultivars are grown throughout northern Europe and some maize cultivars produced in Kenya are used in neighbouring African countries.

Cultivar trials are designed to give as accurate a picture as possible of the performance of the cultivars so as to give a sound basis for predicting what will happen when they are used by farmers to produce crops. The trials are laid down with suitable replications; several different designs are possible, but the most usual is the randomised block, or for large numbers of cultivars, the incomplete block (see Cochran and Cox 1955, and Dyke 1974). The size of plot will depend upon the equipment and the resources available. The trials are conducted at as many locations as possible within the area of adaptation of the cultivars, and are repeated in at least two successive seasons before the results are assessed. Trial management follows local farming practice as closely as possible: trials which are conducted with inputs differing widely from those which farmers are capable of using may give results which indicate a performance which will not be achieved in practice; for example, if fertiliser applications to the trials are much greater than the local farmer would use, the cultivars giving the best results in the trials may not perform well when grown with less fertiliser. On the other hand, it is also possible to design a trial management which will identify cultivars suited to a new package of management practices, provided such a package is within the capabilities of the local farmers.

The statistical analysis of the data will depend upon the design of the trials. Generally, individual trials will first be assessed, and subsequently the combined data can be analysed using an appropriate technique.

Yield of produce is usually the most important criterion on which cultivars are selected, but there are other characteristics which must be assessed. Depending on the crop, field characteristics will be more or less important; for example, for a root crop like fodder beet, the ease with which the roots can be lifted from the soil may be important, while for a cereal, resistance to lodging may be vital for a successful crop. Disease resistance and tolerance of natural hazards – flood, drought, frost, etc. – are always important. Finally, the product of the crop must be of the right quality for the purpose for which it is intended; for example, crops for animal fodder are assessed for feeding value and wheat may be assessed for bread-making or barley for malting.

Cultivar identification

In addition to determining the value of a cultivar for cultivation and use, before it is grown for seed on a wide scale we also need to be able to recognise it during multiplication. It is necessary to know when gross admixture, excessive mutation or pollination by undesirable pollen have occurred. To do this, the characteristics which distinguish the cultivar from all others have to be established so that it is possible to identify seed-lots and seed-fields during multiplication as being consistent with the characteristics of the cultivar. The characteristics which distinguish a cultivar should be easily observed. Whereas the characteristics which determine its value may only be determined after extensive trials, those used for identification need to be available as quickly as possible. Although there are some few characteristics which relate to the seed itself, it is rarely possible to identify a cultivar by reference to seed characteristics alone. Usually it is necessary to grow plants from a sample of the seed to determine the identity of the cultivar. Various morphological features of seeds and plants are assessed for different aspects such as colour, shape and presence or absence of hairs. For this purpose stable characteristics are required so that the identity of the cultivar will not be eroded during multiplication from one generation to the next. Descriptors for use in many crops have been prepared by two international bodies in particular: the International Board for Plant Genetic Resources (IBPGR) and the International Union for the Protection of New Varieties of Plants (UPOV) and those which are relevant have been listed in the Bibliography and References.

A more recent development has been the use of electrophoregrams showing the pattern of storage proteins in the seed. In some crops it has been shown that for a majority of cultivars these patterns are unique to individual cultivars. Considerable progress has been made in identifying wheat cultivars in this way and many other crops are being investigated. Electrophoresis, however, can be achieved by many different techniques and this is therefore a rapidly developing field. Given that consistent results can be obtained electrophoresis can provide a rapid means to identify seed-lots as to cultivar and it need not necessarily be too costly when used as a tool for cultivar identification. However, the assessment of cultivar purity may be more costly because of the larger number of seeds which would have to be assessed.

ASSESSMENT OF CULTIVAR PURITY

The assessment of cultivar purity during multiplication is a vital task. The seed producer must be sure that nothing happens during the growing, harvesting, seed-cleaning or subsequent distribution which might have caused mixture or other dilution of the cultivar. There are three ways in which checks can be made:

1. Tests on samples of seed in the laboratory, including seedling assessment on germinated seeds, i.e. laboratory tests.
2. Tests on growing plants in plots sown with samples of seed, i.e. control plot tests.
3. Inspection of growing seed crops, i.e. field inspection.

Laboratory tests

Cultivar tests in the laboratory are usually of limited value, as except for electrophoresis there are few which will identify the individual cultivars. A first classification can be made from visual observation of the seed – shape, colour or other physical features are all used. Generally this will identify the crop to which the seed belongs, and sometimes classes within crops can be identified; for example, wheat cultivars may be classified on seed colours (white, red or rarely other colours) or maize cultivars may have flint or dent seed. Germinated seed can often show additional characteristics, as for example the presence or absence of anthocyanin pigmentation in the coleoptile of rye. For some crops the ploidy can also be used to classify cultivars, for example diploid and tetraploid ryegrass. A second classification may be possible following chemical tests. Electrophoresis has already been mentioned as a potentially valuable tool. Other tests can be made such as for erucic acid or glucosinolate content of oil-seed rape or reaction of wheat seed to phenol.

Pre- and post-control plots

Control-plot tests are commonly used to monitor cultivar purity during a seed-multiplication programme (Fig. 2.1). The disadvantage of the control

Fig. 2.1 Control plots

plot is that the results are not available until the end of the next growing season after the seed was harvested. Thus the seed to which the test relates has usually been used to produce a further crop by the time the results are known. Nevertheless, control plots provide invaluable information when a further generation is being grown for seed; in such instances the test is known as a 'pre-control test' and the results will be available before or about the time that the next seed crop is harvested so that if faults are detected the crop can be rejected as unfit for seed. When the control plot relates to seed which has been used to sow a food or industrial crop, the result will come too late for remedial action; however, such 'post-control test' results are valuable because they show how efficient or otherwise the seed-production system has been in preserving cultivar purity and suggest ways in which the system might be improved.

In the control-plot test comparisons are made with an authentic sample of seed of the cultivar. Such samples, known as 'control' or 'standard' samples are usually obtained from the plant breeder at the time when cultivar descriptions are being prepared. The samples are large enough to serve for several years and are preserved in special stores – usually cold stores where the sample can be stored in closed tins or jars after drying to an appropriate moisture content. The optimum moisture content will vary with the crop, but is generally less than 10 per cent. When the germination of a standard sample starts to fail, a new sample is usually obtained from the originating plant breeder. The new sample should be included in tests together with the old for at least one season to check that it is genuine before it replaces the original standard sample.

In siting control plots, care must be taken to ensure that the field is suitable: there must be no risk of contamination from volunteer plants of the same crop and this requires a rotation planned to clean the field of self-seeded

Fig. 2.2 Spaced plant trials with grasses

plants and weeds. The tests should be designed to give the required precision: where characteristics are determined by observation, a simple layout with all samples of the same cultivar grouped together will provide the best basis for comparison, standard samples being included at appropriate intervals. The plots should be duplicated in another part of the field, or elsewhere, as a check. However, where characteristics have to be measured, a more formal design is required, such as a randomised block. For some species of grasses and fodder legumes it may be necessary to use a spaced plant design for the plots to permit accurate measurements to be made of plant features such as leaf length or leaf width (Fig. 2.2).

In control plots it is usually possible to observe when a whole plot is wrongly named or badly mixed. It may be more difficult to decide whether or not an individual plant should be classed as an off-type; such decisions require experience since they must be based on subjective judgement as to whether the off-type is a genuine genetic variation, or whether a normal variation between plants has been exaggerated by the environment. Small variations generally have to be ignored and only plants which are clearly different from the cultivar are counted as off-types. In plots where off-type numbers can be counted in relation to the total number of plants in a plot, we have to consider the relationship between the number of off-types observed in a plot test, the desired level of purity and the actual level of purity of the seed-lot. Provided the sample of seed used to sow the plot is truly representative of the seed-lot, and the number of plants in the plot is sufficiently large, the off-type plant count in the plot can be used to calculate a probability of the seed-lot being of the desired level of purity. For ease of working, reject numbers can be used to take account of the risks of wrongly accepting or rejecting the seed-lot. Figure 2.3 illustrates the risks involved with four different sample sizes when the desired level of purity is not lower than 1 per

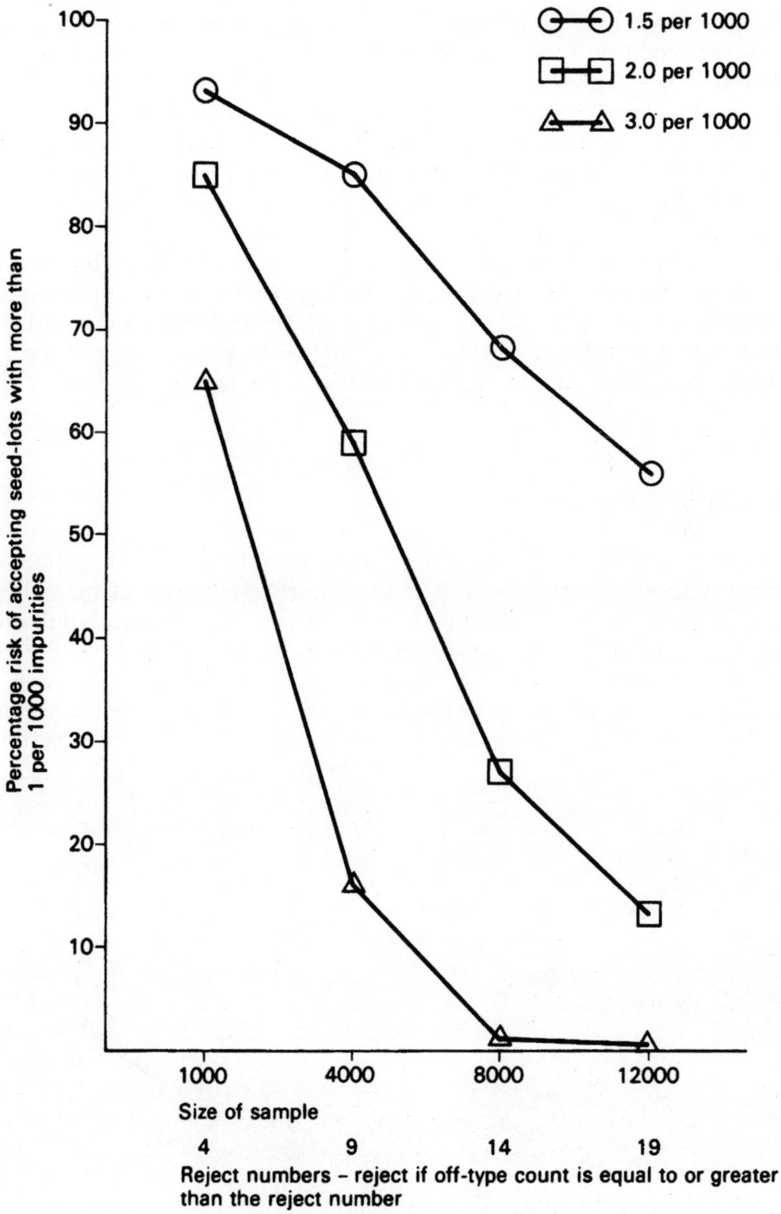

Fig. 2.3 Percentage risk of accepting seed-lots with impurity rates greater than 1 per 1000. (Based on data in OECD, 1982)

1000. In choosing a sample size, the costs and time involved in observing large samples have to be balanced against the risks of reaching a wrong decision. In the examples in Fig. 2.3, the sample size of 4000 gives a reasonably low risk of accepting seed-lots where the true impurity rate is 2 in 1000 (59 per cent) and 3 in 1000 (16 per cent); the risk of accepting lots with a true impurity rate of 1.5 in 1000 (85 per cent) is high, but acceptable in relation to the extra work involved in making a marked improvement –

sample size would have to increase to 12 000 to bring the risk below 60 per cent. As a general rule a sample size of $4n$ can be used, when the desired level of purity is 1 in n; i.e. in the example above a sample size of 4000 when the desired level of purity is 1 in 1000. The reject number of 9 in 4000 is calculated to give no more than a 5 per cent risk of wrongly rejecting seed-lots with fewer than 1 per 1000 impurities. The risk of wrongly accepting a seed-lot is higher than the risk of wrongly rejecting, so as to protect the seed producer with a lower risk of error.

When a randomised block or other design is used and characteristics are measured, comparisons are usually made by analysis of variance of mean plot values for the test samples and the control samples. The convention normally followed is that a difference which is significant at the 1 per cent level of probability ($p = 0.01$) should be regarded as substantiated.

Seed-crop inspection

Field inspection of growing seed crops serves several purposes, but the most important is to check that the crop shows the characteristics of the cultivar which it purports to be. Seed crops may be inspected frequently while they are growing, but at least one inspection has to be timed to allow the best opportunity to assess trueness to cultivar. With most crops this will be during the flowering period or immediately before dehiscence of the anthers;

Fig. 2.4 A well-isolated seed crop

in some crops an earlier or a later visit may be desirable. Other points which are usually checked at field inspection are:

- Was the previous cropping history of the field such that there is a low risk of any undesirable volunteer plants which might contaminate the harvested seed?
- Is the seed crop sufficiently isolated from other crops to reduce the risk of mixture at harvest or pollination by undesirable pollen? (Fig. 2.4). In some instances isolation may also be needed to reduce the risk of infection by seed-borne diseases.
- Is the seed crop reasonably free from seed-borne diseases?
- Is the seed crop reasonably free from weeds and other crop plants especially those whose seeds may be difficult to separate from the crop seed after harvest?
- For hybrid cultivars, is the proportion of male to female plants satisfactory and have the female plants been effectively emasculated, either physically or by genetic control? Note, however, that the latter point cannot easily be checked when self-incompatibility is being used.

The technique of field inspection differs in detail depending upon the particular features of the crop, but there should be a pattern to each inspection which ensures that all important points are covered.

Fig. 2.5 Inspecting a maize seed crop

The person making the inspection should be provided with all information about the seed crop. This will include a description of the cultivar and the history of the seed used to sow the crop, with all relevant results from previous inspections or control plots either in the current or possibly also preceding years. The cropping history of the field should be obtained for the past 3–5 years.

If the person making the inspection is someone other than the seed grower, the first task on arrival at the farm is to check over the field details with the seed grower and to ensure that the correct field is located (Fig. 2.5).

On arrival at the field, the first task is to check that the crop as a whole is consistent with the characteristics of the cultivar given in the description. This is usually done by walking a short distance into the seed crop and examining a reasonable number f plants. The actual number to be examined in each case will depend on the complexity of the distinguishing characteristics (whether easily observed or not) and the variability between plants. Thus it will normally be necessary to look at a larger number of plants in a cross-fertilised species than in a self-fertilised.

The next procedure is to examine the seed crop as a whole by walking round the perimeter; during this time the status of the crop in relation to other crop species, weeds and seed-borne diseases can be assessed and the isolation distances checked. Particular attention should be paid to areas in the crop which may have been contaminated, for example gateways where farm transport may have dropped crop seeds; places where sowing started should be located as equipment may have been brought on to the field before being properly cleaned. The final task is to assess cultivar purity. To do this it is necessary to follow a sampling procedure which will focus attention on small areas of the crop for detailed examination. For crops sown in rows it is possible to estimate a plant population per hectare; counts of off-type plants in the sample area can then be related to this population estimate to give a percentage cultivar purity for the crop. For some crops it is not possible to estimate plant population, for example in seed crops sown broadcast or in dense swards of some perennial grass seed crops where it would be too time consuming to disentangle individual plants to make a plant count. In these crops the estimate of cultivar impurities is usually quoted as number of off-type plants per unit area. Figure 2.6 illustrates some sampling procedures which follow patterns effectively to cover the whole field. While the general pattern adopted should be similar to one of those illustrated, the exact location of each sample area should be random in the sense that there should be no conscious selection of areas which appear to be better or worse than the general run of the crop. In practice this may be achieved by taking a predetermined number of steps between each sample area.

In deciding how many sample areas should be examined in a field it is necessary to balance the requirements for statistical accuracy against what it is practical to do within the time available for making inspections. Generally this will involve compromise, and possibly the acceptance of a greater than desirable risk of reaching a wrong decision; usually this risk is biased in favour of accepting a crop which may have a true impurity level greater than the desired standard, i.e. the decision is biased in favour of the seed producer. This can be justified on the grounds that standards for cultivar purity are usually set higher than is strictly necessary for commercial crop production, so that the buyer of the seed is protected in this way.

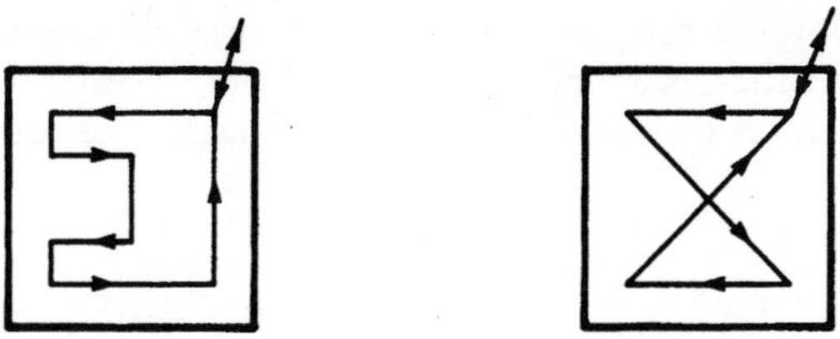

A B

Walking distance approximately the same, but A brings more of the field within 50m of the inspector

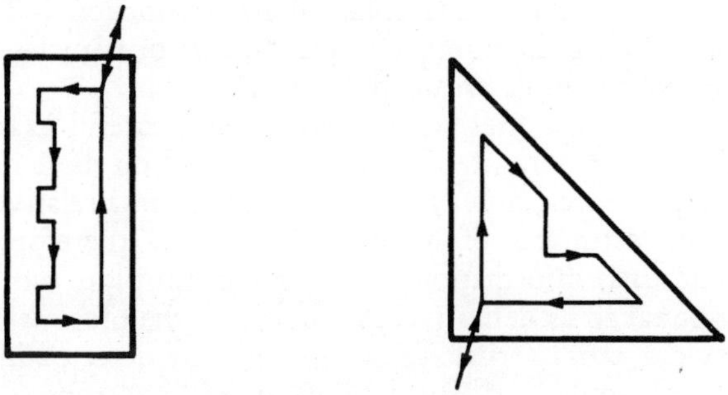

Pattern A can be adapted to other shapes

Fig. 2.6 Patterns for field inspection (approx. 9 ha field)

Inspecting to percentage standards

The sampling procedures will depend in part upon how the genetic quality standards are expressed: as a percentage or as a maximum tolerated number of off-types per unit area.

For standards expressed as a percentage the number of off-types observed in the sample areas has to be related to the plant population. Impurities in a seed crop will differ in the ease with which they can be observed. Obvious impurities are those which differ markedly in colour, shape or tallness from the average of the cultivar; less obvious impurities may only be detected by examination of a particular plant part – for example the hairiness of a leaf or the colour of the seed in the ear, panicle or pod. Larger samples can be examined for obvious impurities than for those which are less obvious. For cereals, sample areas of 20 m^2 each will contain about 10 000 ears or panicles in north-west Europe so a sample 1 m wide by 20 m long walking across the direction of sowing is suitable. For other crops this model can be followed wherever possible, but may have to be adapted for some. For crops sown in wide rows, the sample could be 25 m of row, including the space between rows. The number of sample areas which can be examined is limited by the time available; in cereals, 10 per crop would include a total of about 100 000 ears or panicles, or about 50 000 plants which would be sufficient to give a good estimate of cultivar purity. On the other hand, a crop in wide rows with plants spaced at 50 cm in the row and using a 25 m length would only include 500 plants in 10 samples, so that the estimate would be that much less precise. Using the 'rule of thumb' of $4n$ mentioned in connection with control plots, cereals could be estimated against a theoretical impurity level of 1 in 12 500 and the other crop described at a level of only 1 in 125. Estimates of plant population per hectare can be made using the formula:

$$P = \frac{1\,000\,000\,M}{W}$$

where P = plant population per hectare;
M = mean number of plants per metre length of row; and
W = distance between rows in centimetres

For some crops such as cereals it is usually quicker to count ears or panicles than plants; it then has to be assumed that each plant will produce about the same number of ears so that the counts obtained are proportional. The value of M is obtained by counting the number of plants, ears or panicles in a 1 m length of row within each sample area and taking the mean.

To detect less obvious impurities which may require careful examination of a particular part of each plant, attention can be focused on the 1 m lengths of row within each sample area; however, these will contain a relatively small number of plants and estimates of impurity levels will therefore be less precise. To overcome this, information from the control plots can be used: the plot result may be taken as definitive. Alternatively the plot may be used to relate the less obvious characteristic to one more easily detected in the off-type plant but which would not normally be used; by using the more obvious characteristic the off-types can be counted in the larger sample areas during field inspection. If neither of these solutions is possible, random samples of plants, ears or panicles can be collected for examination in a laboratory, the number being sufficient to determine impurities at the maximum permitted level with reasonable precision.

For ease of working in the field, ready reckoner tables can be produced for appropriate values of M and W to avoid calculating the plant population for each field. Similarly, 'reject numbers' as described for control plots can be calculated for the expected range of populations likely to be encountered in the seed crops to be inspected.

Inspecting to other standards

For standards expressed as a maximum number of impurities per unit area the sampling procedures which will be described are based upon two assumptions: that the impurities are randomly distributed throughout the field, and that counts of impurities have a Poisson distribution. If there are patches of impurities in parts of the field, these assumptions would be invalid; in such cases the patches should be excluded from the samples and inspected separately. In the design of the sampling procedures the risk of making a wrong decision is biased in favour of the seed grower, i.e. there is less risk that a crop will be wrongly rejected and a greater risk that a crop might be wrongly accepted. Two methods are described, designed to test against a standard of not more than 1 impurity per 10 m^2. The methods accept a 20 per cent risk of accepting fields with a true impurity rate of 1.50 impurities/10 m^2 and 10 per cent of accepting fields with 1.05 impurities/10 m^2. For fields where the true impurity rate is less than 1.05 or more than 1.50/m^2 the chance of misclassification is considerably less when not more than 15 counts are made in Method I or 18 in Method II. When 20 counts are made in either method the risks are greater, but this is necessary to avoid continuing the inspection indefinitely.

Method I: Predetermined number of sample counts
In this method, field size is limited to 10 ha and larger fields have to be subdivided for separate inspection. The procedure is to make counts in 15 sample areas each of 10 m^2. If the total number of impurities is 15 or less the field is considered to meet the desired standard (1 impurity/10 m^2). If the total is 22 or more, the standard has been exceeded. If the total lies between 16 and 21 inclusive, 5 more counts are made: if the total of impurities is now 21 or less, the standard is met, but if 22 or more the standard is exceeded.

Method II: Sequential sampling
Although in theory sequential sampling can be valid irrespective of field size, in practice it has been found necessary to make a minimum number of counts

Table 2.1 Method II – sequential sampling. Minimum number of counts for various sizes of field

Size of field (ha)	Number of counts
1 or 2	5
3	7
4	10
5	12
6	14
7	16
8	18
9 or 10	20

Source: OECD (1982).

Table 2.2 Method II – sequential sampling. Acceptance and rejection limits for stated numbers of counts

Number of sample counts	Total impurities	
	equal to or less than: accept	greater than: reject
5	2	13
6	3	14
7	5	15
8	6	17
9	7	18
10	8	19
11	10	21
12	11	22
13	12	23
14	14	24
15	15	26
16	16	27
17	18	28
18	19	30
20	21	22

Source: OECD (1982).

before testing the result against the standard. The procedure is to make the minimum number of counts determined by field size by reference to Table 2.1. The total of impurities in these counts is then assessed against the criteria in Table 2.2. If it falls between the figures for acceptance or rejection, further counts are made with reference to Table 2.2 after each one to see if a decision can be reached. If no decision is reached after eighteen counts, it is necessary to curtail the time spent; two further counts are therefore made and the criteria used in Method I are used (accept if total impurities 21 or less; reject if 22 or more).

Although in theory Method II can save time, in practice this is only true when a majority of crops are well within the standard of 1 impurity/10 m^2. The problem with sequential sampling is that it is difficult to work out a good sampling pattern in the field except one which may involve a lot of walking if many additional samples have to be counted before a decision is reached.

SAFEGUARDING GENETIC QUALITY

In this chapter we have so far discussed methods of estimating genetic quality by reference to the cultivar purity of a seed-lot or a seed crop. This section discusses the measures which a seed grower can take to safeguard the genetic quality of the seed which is produced.

The first and most important measure is to ensure that the seed crop is sown with authentic seed of very high cultivar purity. The normal procedure is for the originating plant breeder to maintain a cultivar. Maintenance procedures vary. The simplest, and probably most effective, is to store a large sample of seed under conditions which will preserve viability for several years – if possible for the expected life of the cultivar. The sample can be authenticated in a control-plot test and portions can be withdrawn each year to start a new cycle of multiplication. An alternative is to reselect frequently, particularly for cultivars of self-fertilised crops; a sample of ears, panicles or pods is taken and seed from each ear, panicle or pod is sown separately so that each individual head or pod row can be closely watched and those not typical of the cultivar discarded. This process can be continued by maintaining each line separately for one or two more generations to permit further selection to be made. In some perennial crops the parent plants can be maintained vegetatively for many years. There are also techniques of micropropagation which can now be used to maintain the parent material of a cultivar. Whatever maintenance procedure is adopted, there will need to be several generations of multiplication; when a sufficient quantity of seed has been produced to provide for seed crops to fulfil market needs in one or two more generations, the seed is usually called 'basic seed'. Basic seed is the last generation produced under the control of the originating plant breeder and thereafter seed producers become responsible for the multiplication of commercial quantities of seed.

Usually the plant breeder stipulates how many further generations it is safe to produce before serious loss of genetic quality occurs. The number of generations will be less for cross-fertilised than for self-fertilised crops.

However, it is often administratively convenient if not technically essential to limit the number of generations after basic seed so as to have an orderly succession of multiplication cycles, each one relating to the plant breeders' maintenance stocks. This concept is called 'generation control'. For a hybrid cultivar there is a formula which must be followed to reconstitute the hybrid each season.

Seed crops can, of course, be rogued. This involves walking through the crop to remove off-type plants. However, this is a costly operation and it is also not very efficient. It is virtually impossible to remove all off-type plants from a crop since they can be quite difficult to see and often are less obvious depending upon the wind or the light – position of sun, extent of cloud, etc. Dense crops are particularly difficult to rogue and those which are lodged are impossible. Nevertheless, seed growers should be encouraged to rogue crops to give the fine tuning to an already reasonably pure crop. It should be understood, however, that roguing will not rectify a crop which already contains many impurities.

Apart from ensuring that the crop is started by sowing the correct seed, other measures to safeguard genetic quality generally require attention to detail: the field must have a satisfactory cropping history so as to minimise the risk of volunteer plants and must be sufficiently isolated; all equipment brought on to the field must be properly cleaned before entering; irrigation water should not contain any seeds; at harvest and subsequently great care must be exercised to prevent mixture occurring with other seed.

During the growing, harvesting and subsequent cleaning and distribution of the seed crop it is important to keep good records. A historical account of the progress of the crop and all operations in the field can be invaluable subsequently in determining where a mistake might have occurred leading to loss of genetic quality.

VIABILITY AND GERMINATION

After genetic quality, the viability of the seed is the next most important aspect of quality. Seed should contain the highest possible proportion which is capable of vigorous germination. In any seed-lot some seed will be dead, and some may produce abnormal seedlings which will not grow into healthy plants. The germination test is designed to determine the germination capacity of a seed-lot (percentage by numbers) by testing a sample in the laboratory. Usually this involves germinating a subsample of some 400 seeds in 4 replicates of 100 seeds each under controlled conditions adapted to the requirements of the crop. Other quicker methods such as tetrazolium staining which detects dead tissue in an embryo have been devised, but are less accurate. Germination tests are undertaken only on pure seed of the crop. Through the efforts of ISTA there is now a wide measure of agreement when tests are conducted in different laboratories on samples from the same seed-lot. However, this depends also on the skill and training of the seed analysts who perform the work.

The ISTA test procedures determine germination capacity. Vigour of germination is also important and various tests have been devised to determine vigour. Generally these involve germinating seed under stress conditions (e.g. at colder than optimum temperature). However, these tests have generally not given sufficiently consistent results for them to be included in the ISTA rules. They may serve to detect samples with exceptionally low vigour, but are less capable of detecting degrees of vigour among a group of reasonably vigorous seed-lots. In practice, therefore, most quality-control measures concentrate on determining germination capacity on the assumption that seed-lots showing a high capacity will also be vigorous.

ANALYTICAL QUALITY

The laboratory purity test determines the proportion of pure seed in the seed-lot by dividing a sample into crop seed, other seed and inert matter. The 'other seed' fraction may contain seed of other crops or of weeds. These seeds should be identified. Some of the 'other crop seeds' may be similar in shape, size and weight to the cultivar seed and so be very difficult to remove. Weed seeds also may be difficult to remove, and some may be particularly objectionable (often termed 'noxious') because they are very difficult to eradicate in the field, or they may be poisonous. The results of the purity tests are expressed as percentage by weight.

An additional test of analytical purity is to determine the numbers of seeds of particular species occurring in a given weight of sample. The normal purity test uses a small subsample, the weight of which is adjusted for each crop so that it will contain approximately 2500 seeds. For determination of particular kinds of seed per unit weight samples of about ten times this size are needed. The purpose of the seeds per unit weight estimation is to take account of particular seeds which are dangerous to the user of the seed. If a farm does not contain any plants of a particular noxious weed, the farmer will wish to ensure that seed brought on to the farm does not contain any of its seeds. The normal purity test uses too small a sample to give a reasonable assurance on this point, but the search of a larger sample will give a better estimate. The ISTA rules contain details of the purity test and the tests to determine specified seed numbers per unit weight. The accuracy of the results of these tests, as with the germination test, depends mainly on the skill of the seed analyst. Equally, the person drawing the sample for analysis must be skilled and must ensure that a representative sample is obtained: ISTA rules prescribe methods for sampling, covering the maximum size of seed-lot to be represented by one sample, the number of subsamples which should be drawn to make up the sample and the methods by which the subsamples should be obtained.

HEALTH QUALITY

Some diseases or pests are seed-borne and unhealthy seed can thus provide foci of infection when sown. Infection may be by fungi, by virus or micro-organisms, or insect pests may be living in or on the seed. The actual amount of infection may be quite small, but because many diseases or pests are able to multiply quickly as the sown seed germinates, small infections can nevertheless cause extensive damage to the crop.

Seed health testing has been a subject of study in ISTA of recent years and procedures have been developed for many of the more important seed-borne diseases. These procedures can only be followed by personnel who have been given a basic training, and some require special equipment. The laboratory test uses relatively small numbers of seeds, and therefore it is difficult to detect small numbers of infected seed in a seed-lot. A nil result in a test does not necessarily mean that there is no infection. Laboratory test results, therefore, have to be interpreted with care. It is also possible to monitor seed health in the seed crop during field inspection and in the control plots. However, these results will be historical and it is by no means certain that the harvested seed will be more or less infected: conditions for the spread of the disease in the seed crop may have been good, in which case infection in the next crop may be greater, or the reverse may happen so that the seed has less infection than might have been expected. Hewett (1981) has pointed out that there are differences in the way in which a pest or disease spreads. Some pathogens are relatively uncomplicated being dependent on the host crop plant and only able to spread infection when conditions are favourable. Others can exist on various host plants or crop residues as well as the host crop plant and are more difficult to control since they are capable of reinfecting a crop which has been treated successfully against the pathogen.

For the seed producer, the important point is to start with a healthy seed stock, and subsequently to take all possible precautions by crop rotation and isolation of seed crops to avoid infection. Many diseases can be treated; however, treatment is often expensive and only possible on relatively small quantities. For this reason basic seed (or even pre-basic seed) is often treated and subsequently the following seed generations are maintained as disease free as possible by practising field and crop hygiene and providing isolation from possible sources of infection.

PHYSICAL QUALITY

For the farmer who buys seed appearance may be all important, because it is the only criterion which can be observed. Although it is true that many seed samples which have a poor appearance will grow satisfactorily when sown, it is usual to prefer larger, plumper, bright-looking seed. In general, this plumper seed may be expected to perform better.

Uniformity of size is also important. Generally a uniformly sized seed-lot will give a more uniform crop than one which has not been size graded. For some crops, seed is sold in different size grades, particularly when the crops are sown with seeds spaced at relatively wide intervals; for example beet, maize and some *Brassica* seed. When seed is pelleted before sowing, size grading is essential.

Mechanical damage may be caused by careless handling of seed during or after harvest. Cracked or broken seed can normally be removed during seed cleaning, but some seeds can also be damaged by quite minor hair cracks in the seed coat; such seeds are difficult to remove from the seed-lot and are likely not to germinate satisfactorily. Mechanical damage provides opportunity for the ingress of pathogens which may destroy the seed in storage. Seed thus has to be handled with care. Conveyance systems which may be quite satisfactory for grain may not be suitable for some seeds.

STORAGE QUALITY

The moisture content of the seed generally governs the length of time for which it can be stored, coupled with the temperature within the store. Harrington and Douglas (1970) devised two simple rules:

1. For every decrease of 1 per cent in seed moisture content the life of the seed is doubled.
2. For every decrease of 5 °C in storage temperature the life of the seed is doubled.

These rules operate with seed moisture contents between 14 and 5 per cent, and with temperatures from 50 to 0° C and are independent of one another. When moisture content is above 14 per cent, moulds can grow on the seed, and when it is above 18 per cent, heating can occur. In general, seed between 10 cent and 14 per cent moisture content will maintain viability until the next sowing season (i.e. between 6 months and 1 year) while seed below 10 per cent can be kept for longer periods. Low temperatures prolong the safe storage period. Insect activity is greatest between 21 and 27 °C and therefore storage below 20 °C should be provided wherever possible. The relative humidity (RH) of the air in the store is also important since seed dried to a low moisture content will eventually pick up moisture from the air if the RH is high. From the point of view of storage quality, however, it is the moisture content of the seed which is the all-important factor, and it is necessary to make checks on moisture content while seed is in store. Moisture content is estimated on a sample of the seed, and the ISTA rules provide suitable methods. Samples drawn for this purpose should be transported in moisture-proof containers such as a plastic bag if the test cannot be done on the spot. The most exact method of determination is to dry to constant weight in an oven, but various quicker methods (for example, using electro-conductivity or capacitance or infra-red) are available; speed will usually cause some loss of accuracy.

Storage quality is also influenced by other factors. Seed which has been damaged by rough handling, or which contains a high proportion of rubbish – broken pieces of straw or leaf – will not store well. The normal practice, therefore, is to pre-clean seed before putting it into store to await processing. This operation removes broken seed and other trash. Weathering in the field before harvest can also cause storage quality to deteriorate. In some conditions, weathering can cause germination to start, and subsequent drying will suspend this activity. When this happens germination capacity can deteriorate rapidly in store although it can be quite difficult to detect this condition; germination tests soon after harvest may well give satisfactory results.

QUALITY CONTROL DURING SEED PRODUCTION

To produce good-quality seed we need to have a system for quality control which will monitor all aspects of production. In this seed is no different from any other product. The food on supermarket shelves, for example, is all subject to quality control while it is being processed and packed, from the moment it is delivered as raw material, to the time when it leaves the processing plant. However, there is a difference between such a product and seed. For most products sold in shops it is possible to take samples at any time and to make quality tests which can give results within a short time – at most a week or two. Therefore any faulty products discovered can normally be withdrawn before they are offered for sale. With seed, however, although laboratory tests for germination, purity, moisture content, etc. can be done quickly, tests for cultivar purity normally take an entire growing season to complete. Thus the control of the vital aspect of genetic quality relies heavily on control of the production processes, and the seed is generally sown before genetic quality can be verified. In some cases even the full germination test may be impracticable for the processed seed which is actually sown; for some autumn-sown crops the interval between harvest and sowing is so short that results from tests of uncleaned, untreated seed may have to be used. The seed purchaser, therefore, relies to a great extent upon the integrity and conscientiousness of the seed producer, and this in turn calls for a clear understanding of responsibilities throughout the entire seed-production process from the plant breeder to the final commercial seed quantities.

The usual custom on responsibility is to regard the owner of the seed or seed crop for the time being as responsible for safeguarding quality. During the production of basic seed the plant breeder is usually held to be responsible, and subsequently as ownership changes with the production of further generations so this responsibility is passed on. As soon as a seed grower starts to sow basic seed that person accepts responsibility for safeguarding quality until the harvested seed leaves the farm. The new owner may be a seed merchant with a seed-cleaning and treatment plant in which the seed-lot must be kept separate and any mixture avoided; storage conditions must be suitable until the seed is passed on. The next stage may be another seed grower, a distribution agent or a commercial crop grower, but responsibility for quality will rest with the owner at each stage.

To fulfil their responsibilities these people will need thoroughly to understand the principles of quality control described earlier, and to ensure that members of their staff receive adequate training to enable them to fulfil their parts. They must also have access to seed laboratories where tests can be made. Many seed companies employ technically qualified staff who are able to inspect seed crops and undertake laboratory tests; some may also undertake plant breeding and as a consequence will have a system of variety trials.

However, seed is considered to be such an important input in the agricultural industry that in most countries the government also takes an active part in establishing good-quality supplies.

The extent to which a government intervenes to ensure that seed is of good quality varies in different countries. In most countries there is a national agricultural research organisation financed by government which includes a plant-breeding capability. These organisations may include a cultivar-testing capability or be complemented by a separate cultivar-testing organisation. In either case, official cultivar testing, either compulsory under the law or voluntary, usually brings together the products of both private and publicly financed plant breeding for final assessment. The way in which these organisations are set up can vary from a department in the ministry of agriculture to semi-autonomous institutions, sometimes under universities. Basic seed of publicly financed cultivars is produced by the national research organisation, either in the plant-breeding department or in a separate department set up for that purpose with the plant breeders giving advice in relation to particular cultivars. Basic seed of cultivars bred by private plant breeders is normally produced and marketed by them, often within a larger seed company.

Plant breeders' rights

In some countries, private plant breeders benefit from a system of plant breeders' rights which is essentially similar to an author's copyright or an industrial patent. Judgement as to whether or not a new cultivar qualifies for rights is made by the patent office in a country or a parallel organisation dealing only with cultivars. The criteria on which rights are granted are regulated by the International Convention for the Protection of New Varieties of Plants (UPOV 1978), which is coordinated by UPOV with headquarters in Geneva in the offices of the World Intellectual Property Organisation; UPOV is able to give advice on both the technical and the legal aspects of plant breeders' rights. The registration of cultivars for plant breeders' rights provides an authentic description of each cultivar which can be used for quality-control purposes. If there are no plant breeders' rights in a country, these descriptions are usually provided by the plant breeders or by the cultivar-testing organisation.

Seed certification

There are several ways in which quality control can be organised after the basic seed stage, but there are two main aspects to be considered. First there is quality control during production up to the time when the seed is ready for sale; and secondly there is the control of seed during marketing. Quality-control measures during seed production can be undertaken by the seed producers themselves, and most seed producers will wish to take an active part in ensuring that the seed they produce is of good quality. However, in many countries the government provides a service with quality-control measures grouped together in a 'seed certification scheme'. These schemes may be voluntary, when the seed producer can choose whether or not seed crops should be entered, or it may be a legal requirement that all seed producers enter their crops so that the sale of seed which has not been officially certified is prohibited.

Internationally, the OECD seed schemes include some thirty-eight participating countries. These schemes cover herbage and oil-seeds, cereals, beet, maize, subterranean clover and also vegetables and forest trees. They differ in detail, but all are based upon the following main principles which are basic to any seed-certification scheme:

1. Cultivars are admitted to the schemes only after official trials have shown them to be valuable for cultivation and use and to have distinguishing characteristics.
2. Each cultivar is to be maintained by the plant breeder responsible for its production, who must also ensure that seed is available each year to provide continuity in the scheme; the authority in each country responsible for supervising the scheme (designated authority) makes checks in the laboratory and in control plots to ensure that the seed produced by the plant breeder (or under the plant breeder's control) is satisfactory for all aspects of quality.
3. Fields for the production of certified seed are inspected by representatives of the designated authority and must conform to published standards.
4. Seed harvested from approved fields should be cleaned and treated as appropriate and is subsequently sealed in labelled containers and sampled. The seed producer is responsible for making laboratory tests for germination and purity. A proportion of the samples is checked by the designated authority in laboratory tests and control plots. Where further generations of certified seed are to be grown, checking the samples is mandatory.
5. All samples are drawn by authorised representatives of the designated authority. Where appropriate, ISTA rules are followed in making the check tests.
6. The seed produced by the plant breeder is known as 'basic seed'. Preceding generations are known as 'pre-basic seed' and may also be supervised by the designated authority. The generations preceding basic seed are limited by agreement between the plant breeder and the designated authority.
7. The certified seed is designated as first, second or other generation after basic seed. The number of generations which may be produced is limited by agreement between the plant breeder and the designated authority.

These principles outline the essential points which any seed-certification scheme should contain. Additionally, schemes will contain requirements and minimum quality standards which are specific to the particular crops included. These requirements and standards cover the aspects of quality discussed earlier in this chapter, and also provide details of how the various field inspections, control plots and laboratory tests are to be conducted.

It will be evident from the foregoing that seed certification, so far as genetic quality is concerned, is primarily concerned with the way in which seed crops are grown and the harvested seed is handled. The various checks that are made are timed to take place at vital points in the production process and it is assumed that if approval is warranted at these points, the quality of the seed should be good. This emphasises the need for it to be clearly-recognised that responsibility for the seed lies with the producer; the certifying authority can only certify that to the best of its knowledge the various rules have been followed, but the producer must ensure that the rules are not broken during the majority of the time it takes to produce the seed when the authority cannot be present.

Quality control during marketing

Control of seed quality by the government during marketing is rather different from seed certification. Whereas seed certification may be regarded as a service to seed producers, providing them with expert advice on quality matters, control during marketing may be likened to consumer protection legislation. Minimum standards of quality are specified for seed offered for sale, and seed sellers are required to ensure that seed is at least of this quality. Another possibility is the so-called 'truth-in-labelling' concept which requires the seller to make certain statements about the quality of the seed on offer but does not set minimum standards. In both cases the government is able to take seed samples in the market-place and to make independent quality tests on these. Obviously it is not possible for samples to be taken from all seed, and the system therefore relies upon random checks being made by government inspectors at appropriate times of the year; a check of about 10 per cent of the seed on offer is usually regarded as satisfactory. If these tests show that seed is being offered for sale below the required standards, the person responsible can be prosecuted in the courts and if found guilty would be liable for a penalty. For quality tests which can be made in a laboratory, a reasonable time-scale is possible, and tests may be complete in time to prevent poor-quality seeds from being sold. However, for the important aspect of genetic quality when a control-plot test is needed, the time-scale is such that action cannot be taken to prevent sale.

SEED LEGISLATION

In most countries where seed production has been developed on some scale there is seed legislation. Because it is generally of a technical nature, the law on seeds is specific. Normally the law regulates the quality of the seed which may be offered for sale and there are two main alternative ways in which this is achieved.

In the 'comprehensive regulatory system' the law limits the sale of seed to a particular minimum standard of quality. There is a register of cultivars, and only certified seed of registered cultivars may be offered for sale. Often seed producers and seed traders must also be registered to stay in business.

The alternative is the 'truth-in-labelling' system, under which the seller is required to give certain prescribed information to the buyer about seed on offer, but there is no restriction on the quality of the seed.

Most seed laws are in general terms, setting out the principles which should be observed and designating an authority with power to fill in the details by regulation. This enables the law to function in the changing circumstances from season to season without the necessity to present amendments to the full legislature which would be very time consuming. A seeds act will usually contain the following essential elements:

1. Usually the minister responsible for agriculture is given ultimate responsibility, but with power to designate organisations to look after one or more of the different points which the law seeks to regulate. When standards have to be set the minister usually promulgates the regulations, but acting on the advice of the appropriate organisation. Organisations may be designated for any of the following tasks; for a 'truth-in-labelling' system, only the last is essential:
   (a) cultivar testing and registration;
   (b) seed quality control which may include seed certification and seed testing and maintaining a register of seed producers;
   (c) market control which may include maintaining a register of seed sellers.
2. The scope of the law is defined in terms of the crops to be covered, and definitions are given which should give guidance as to whether or not a particular commodity falls within the law. For example, 'seed' may be held to include vegetative plant parts used for propagation, or these may be excluded.
3. An outline is given of the quality traits which are to be regulated and for which standards are to be set. These standards may apply during seed production, as in a seed-certification scheme, and at point of sale. Only the latter would be needed for 'truth-in-labelling'.
4. Unlawful acts are specified and penalties prescribed. An appeal procedure may also be outlined. Tolerances allowed between test results may be quoted.
5. The qualifications of inspectors or other official personnel having executive powers are specified and their duties and power defined.
6. Any special points affecting the import or export of seed are given.
7. Exemptions from the regulations are given. For example, farmers who save seed for their own use are usually exempted from the control.

The main purpose of a seed law is to create conditions in which the genuine trader can prosper and will be protected from the activities of the unscrupulous entrepreneur. In this way the farmers who buy seed are also protected. The various organisations do not function solely as regulatory bodies, but also provide centres for research and advice on seed production and quality-control matters.

3 PRINCIPLES OF SEED GROWING

PLANNING FOR THE CROP

To grow a good seed crop needs careful preparation. While yield of seed is obviously of great importance, quality is equally important. A high yield of poor quality is rarely a financial success. To produce a high yield of good-quality seed requires careful forward planning, sometimes over several years.

As we have seen in the previous chapter, a field has to be prepared for a seed crop several seasons in advance. The preceding crops have to be such that the unwanted volunteer plants, weeds and seed-borne diseases can be eliminated or reduced to a minimum. In addition to the actual crops which are grown in the preceding years, it may also be desirable to take special measures in preparation for a seed crop which might otherwise not be warranted – for example, additional weed-control measures by chemical or other means to eliminate a particularly undesirable weed such as wild oat (*Avena fatua* L.).

Forward planning is also necessary in relation to isolation. For cross-fertilised crops in particular, the cropping of surrounding fields must be planned in advance so that no problems will arise. Sometimes it is advisable, particularly with perennial crops such as grasses, to grow only one cultivar of a crop species on a farm both as seed crops and for any additional crops which may be grown; careful zoning will then be needed on the farm when it is desirable to change from one cultivar to another. Such a farm zoning scheme may also be extended to a district scheme, requiring the agreement and co-operation of several farmers. In some countries, district zoning schemes can be established under the seeds legislation, restricting the freedom of action of farmers within a defined zone so as to provide for the needs for seed growing; this may be extended to include private gardens, for example, in relation to certain brassicas which some gardeners might not destroy before flowering thus posing a danger to brassica seed crops in the vicinity. Where no such restrictions exist, the seed grower will have to establish good relations with people in the neighbourhood so that isolation problems are reduced to a minimum in advance of the seed crop.

Plans also have to be made for harvesting, cleaning, storing and treating the seed and for marketing. Speculative seed growing is rapidly diminishing and most seed crops are now grown with a particular market in view, or to provide for a farmer's own needs. Post-harvest operations may be undertaken by the seed grower, in which case the farm should be equipped with suitable seed-cleaning and treatment machinery and suitable storage. Alternatively the harvested seed may be passed on to a seed-trading organisation for processing and storage, and which would subsequently undertake the marketing. Many seed crops are grown on a contract basis for a seed-trading organisation which may be wholesale, retail or both, and may also have a plant-breeding capability.

A check-list of points to be covered before establishing a seed crop would be as follows:

1. Is the cropping history of the proposed field satisfactory for the intended crop?
2. Has the intended cropping of the surrounding fields been checked to ensure that there will be sufficient isolation?
3. For cross-fertilised crops, are the local people aware of the importance of preventing possible donors of unwanted pollen from flowering?
4. Is the proposed field now in a satisfactory state to grow a good, clean crop?
5. Have arrangements been made to deal promptly with the harvested seed, to ensure that it goes into store in good condition?

GROWING THE CROP

A seed crop has to be well grown. The methods used to grow a good crop for food or industrial use generally apply also to the seed crop, although there may be variations. For example, it is sometimes desirable to use a plant spacing for a seed crop different from that which would be used normally, or the timing of some operations may have to be adjusted for a seed crop, and excessive use of fertiliser should be avoided. But crops which are poorly grown, weedy or badly lodged are of little value for seed. As a general rule, therefore, the seed grower should follow the best husbandry practices for the crop, introducing only those variations which have been shown to be beneficial for the production of good-quality seed. Where such variations are specific to the crop, they are discussed in the later sections. There are, however, some general principles by which the seed grower should be guided.

To get the best results from any crop requires attention to detail. Management has to ensure the timely undertaking of all operations from sowing to harvest, including remedial measures which may be necessary to combat weeds, pests or diseases. The seed grower, however, has to be alive to additional details which the ordinary farmer might be able to ignore. Mostly these concern the quality-control measures we discussed in the previous chapter.

First, a seed grower must ensure that everything brought into the field is free from any seed which might contaminate the seed crop. Cultivation and other equipment should be cleaned of any soil, and this should be done well away from the field where the seed crop is to be grown. Equipment for sowing seed, especially, requires thorough cleaning before entering the field for sowing.

Second, the seed to be used to sow the seed crop should be checked carefully. It will normally arrive in sealed and labelled bags and the seals should not be broken until the seed is in the field, ready for sowing. The store in which the seed is kept prior to sowing and the cart or trailer used to move it from store to field should be carefully cleaned before use. When the seed is sown, the labels should be preserved to provide a record of what was actually used. Any seed surplus to requirements should be carefully labelled and returned to a clean store so that it can be used later if required.

Third, a seed grower should check other inputs to be sure that they do not carry any unwanted seed to the crop. Fertiliser will usually be satisfactory when delivered in plastic or similar bags, but occasionally it may be transported in vehicles which have not been thoroughly swept of grain. Farmyard manure usually contains seeds and should only be used in seed-production fields when it is well rotted, and preferably not in the year in which the seed crop is grown.

Fourth, isolation should be safeguarded at all times. This may involve removing or cutting back plants in neighbouring fields before they flower. For some crops it may also be necessary to keep an eye on hedgerows, roadside verges and gardens in the vicinity; for example, many grasses grow alongside roads and can form a reservoir of undesirable pollen near a grass-seed crop; in a garden, many vegetables can flower and cross-pollinate with adjacent seed crops (red beet with fodder beet or cabbage with kale). In some cases it may be possible to arrange isolation in time rather than space, but to ensure true isolation there generally has to be a 30-day interval between flowering in one crop and flowering in the other from which isolation is desired.

Fifth, a seed grower must observe a seed crop regularly and be prepared to take corrective action when problems arise. Weed control is essential and requires particular attention when the weeds concerned have seeds which are difficult to remove from the crop seed; for example, wild oat (*Avena fatua* L.) in a wheat seed crop. Hand roguing of such weeds is often necessary rather than using chemicals when the weed numbers involved are relatively small. Caution should be exercised in the use of hormone-type weed-killers as they can cause distortion in the crop plants; this can lead to difficulties in assessing cultivar purity during field inspection. Disease control is also essential. If seed-borne diseases can be controlled this is particularly important, but other diseases may cause seed to be poorly filled when the crop is harvested; chemical control of leaf diseases is therefore desirable in many seed crops.

Finally, the key to successful seed growing is timeliness. It is essential to be 'on top of the job'. A seed grower needs a keen eye for detail and must be prepared to take advantage of every opportunity to improve the condition of the crop. Seed crops, whether large or small, have to be inspected regularly, and more frequently at the critical growth stages for the crop concerned when action may be needed – establishment, weed, pest and disease control, isolation and when crop inspection to check cultivar purity is possible. Careful and timely attention to these details will safeguard the crops to be grown from the seed in the future.

HARVESTING

Harvest and the aftercare of the seed is most critical in relation to viability of the seed. Time of harvest and method both have to be arranged so as to give the best chance that the resulting seed will have good germination capacity. In addition, the seed grower should ensure that no unwanted mixtures occur and that the seed is stored in good conditions – in short, attending to all the details which will ensure good quality as well as satisfactory yields.

Time of harvest has to be adjusted to ensure that the seed is sufficiently mature, usually indicated by moisture content and appearance. If harvest is too early, the seed may become shrivelled on drying, and although germination may be satisfactory at the start, it can deteriorate in store; cleaning losses are also liable to be high. On the other hand, if harvest is delayed too long, some seed may be shed and the remainder be so dry that it is liable to be damaged during threshing and subsequent handling. Safe moisture content at which to begin harvest varies with the crop: wheat can normally be harvested when moisture content is 20 per cent or below; tetraploid ryegrass, on the other hand, can be harvested at 40 per cent or below; moisture contents below about 10 per cent for wheat or 35 per cent for ryegrass can pose problems with shedding or other losses. Apart from rubbing out a sample of the seed in the field and testing the moisture content, hardness can be judged by pressing individual seeds with a thumb-nail: seed should usually be at least doughy before harvest, but not so hard that it cannot be marked with the thumb-nail.

Appearance of the plants or seed can also indicate time to harvest. As the plant loses its green colour, so maturity approaches. The seed changes colour as it ripens. In some crops these changes have to proceed a long way before the crop is ready for harvest: bean plants (*Vicia faba* L.) for example, usually turn black and lose their leaves before the seed is ripe; some other crops, such as the grasses, may still have traces of green colour in the stems when the seed is ripe. There are also crops which do not ripen uniformly and judgement is required to obtain the majority of the seed in good condition; clover is difficult in this respect because it flowers and sets seed over a long period so that ripening is also prolonged. Where appropriate, an indication of how to judge ripeness is given in the later sections where individual crops are discussed.

Methods of harvesting

There are two main ways in which a seed crop can be harvested: the plants may be cut and allowed to dry in the field or be dried by other means before threshing; or the seed may be removed immediately from the plants and taken from the field for further processing.

The first method allows time for the seed to mature further on the plants before threshing; harvest can thus start earlier, but takes longer to complete. It has a particular advantage for those crops which flower and set seed over an extended period. The second is quicker and usually requires less labour once harvesting begins, but crops are generally more vulnerable to adverse weather the more mature they become (greater risk of shedding or sprouting in the ear). For some crops, chemical desiccation can be used to bring forward the time of harvest and to reduce the amount of green growth in the crop.

Whichever method is used there is the possibility of using simple hand tools when production is small scale and labour plentiful (for example in small village communities). The plants are cut by a tool such as a sickle or scythe and may be left loose to mature, or tied in bundles or sheaves and placed in stacks before being moved to a threshing floor where the seed can be beaten out and winnowed from the straw and chaff. Alternatively, the cut plants can be moved from the field to dry in the sun on a prepared floor or flat roof; this allows for the plants to be covered when there is risk of rain. When the latter method is used it is possible to gather only the seed heads, leaving the stalks in the field for harvesting as forage or for grazing. Hand harvesting may also be practised by plant breeders during the early stages of multiplication, but threshing is then usually accomplished with a small-size thresher.

When crops are cut mechanically for threshing later the choice lies between a binder and a reaper or special windrower (Fig. 3.1). The sheaves from a binder have to be stooked by hand or the binder may be used as a windrower or swather. Windrows may also be collected and placed on tripods or racks to aid drying, but more usually are left on the field. Care must be taken when handling the cut material to avoid excessive seed shedding, and for moving windrows which have been badly affected by rain it is best to use a draper pick-up to lift the windrow and drop it gently behind the machine.

Fig. 3.1 Self-propelled swather (Photograph: Shelbourne Reynolds Engineering Ltd)

Fig. 3.2 Pick-up header for combine (Photograph: Shelbourne Reynolds Engineering Ltd)

Crops which have been stooked, placed on racks or windrowed can be collected from the field and taken to a stationary thresher. However, seed loss during handling is likely to be high and it is more usual to use a combine. This may be taken from stook to stook in the field, the knife and reel being removed and the sheaves being fed in by hand, or for windrows the crop is picked up using a pick-up attachment on the combine header (Fig. 3.2). Picking up a windrow without an attachment by undercutting the windrows usually loses much seed by shedding.

For many crops direct combining is possible, so that the seed is immediately secured and can be taken into store for further conditioning. Crops which ripen evenly are particularly suited to this method and it has advantages for some lodged crops where seed may be shed, but held within the laid plants; for example, crops of ryegrass (*Lolium* spp.) which usually lodge soon after the seed has set can best be handled in this way.

For larger areas, the choice must lie between windrowing and picking up with a combine or direct combining. The former is preferred for crops which do not ripen uniformly or where there is risk of weather causing excessive shedding in a standing crop. The latter has the advantage of securing the seed immediately if conditions are right, and is less demanding of labour.

Setting equipment

Using machinery, and even hand operations can cause seed loss. Plants which are handled roughly or equipment which is badly set will cause excessive shedding or damage to the seed. The seed grower has to balance the need for a slower than normal speed of operation against the need to take advantage of good harvest weather to get the seed safely into store as soon as possible. In some instances it may only be possible to work for limited periods in the day – some grasses are best harvested when the morning dew dampens the crop and prevents excessive seed shedding at the header; others will only thresh satisfactorily if allowed a period of alternate wetting and drying in the field.

Careful setting of machinery pays dividends. When cutting the crop, either windrowing or direct combining, height of cut should be carefully adjusted; it is generally advisable to leave as much stubble as possible. If a crop is cut too low there will be an excessive amount of straw to handle during threshing, but if cut too high some seed heads or pods may be lost. Excessive forward speed can cause loss of seed at the cutter bar. Reel setting should be adjusted to suit the crop: for lodged crops the reel usually has to be set further forward and lower (except where lifters are fitted) and a slower reel speed is also advisable.

Probably the most critical setting for a seed crop, however, is the cylinder speed and concave clearance when threshing, whether this be a combine or a stationary thresher. When dealing with a seed crop the slowest cylinder speed and widest concave clearance commensurate with good threshing should be used. The aim must be to avoid damage to the seed by giving it too hard a knock, and to avoid threshing out immature seed or breaking up the straw and green leaf too much. On the other hand, it is essential to ensure that all the good seed is removed. When combining, settings may require to be changed during the day as the crop becomes drier after a morning dew.

After the cylinder adjustments, the next most important are to the cleaning shoe. The top sieve serves to separate the larger pieces of broken straw and debris from the seed, and return them to the ground. The lower sieve allows the seed to pass through and passes the retained material, mostly unthreshed heads, to the tailings auger which feeds them back to the cylinder via the tailings elevator. Adjustment should ensure that there is not too much material returned which might overload the cylinder. Both sieves are normally adjustable or replaceable so that a variety of shapes and sizes of holes is available. For small-seeded crops such as clovers and brassicas a round hole is preferable to a slot. The sieves are operating in a blast of air from a fan designed to lift the light material from the top sieve and blow it out of the back of the machine. The strength of this air-blast can be adjusted, usually by shutters which regulate the amount of air reaching the fan. The airflow has to be adjusted so that the sieves are kept clean, but should not be so strong as to blow good seed over the back along with the chaff and other debris.

The good seed passing through the cleaning shoe is delivered to the clean-seed auger and thence via an elevator either to be bagged off or, on a combine, to a bulk tank. In stationary threshers and some combines there may be a secondary cleaning in a rotary sieve or similar before the seed is finally delivered.

For seed crops it is particularly important to ensure that all settings are correct. There are also two other points which the seed grower has to watch. First, the machine has to be carefully cleaned before entering a seed crop, particularly when changing from one cultivar to another of the same crop. This is a long job as combines or threshers are not normally designed for ease of cleaning. Running the machine empty for 15–20 minutes will remove the bulk of material, but will leave seed lodged in various parts, particularly at the ends of shafts, around the cylinder and concave, and in the bottoms of augers or elevators. For the regular seed grower it is desirable to cut access doors in the machine and to make hinged bottoms to augers and elevators. An industrial vacuum cleaner and a compressor are indispensable tools when cleaning these machines. A small quantity of the first seed through the machine after cleaning should be discarded.

Secondly, the seed grower wishing to use a combine or thresher for a small-seeded crop – clover, brassicas, linseed – will need to pay particular attention to places in the framework and bodywork of the machine where seed might escape. These machines are normally designed for larger seeds such as cereals, and small seeds can be lost through joints which would usually cause no trouble. Plasticine or other flexible material can be used to avoid seed losses in this way.

The modern self-propelled combine is usually designed for bulk handling of cereal crops and it is increasingly difficult to obtain smaller machines which are suitable for use in smaller areas and specialist seed crops. Machines with delivery to a bagging platform are desirable in preference to a bulk tank when handling small quantities of particularly valuable seed.

Specialist equipment

For some crops specialist equipment has been designed. Maize is best handled by a corn-picking machine, which removes the ears from the plant and dehusks them. Some machines also shell the ears, but for seed it is generally desirable to do this as a second operation in a stationary sheller; this permits the seed to be conditioned on the ear before shelling, either in a drying cage or in a drier, and causes less damage. Combines which have been adapted for maize harvest are not recommended as they can cause damage to the seed.

For rice, combines are usually fitted with special tyres to enable them to travel in wet conditions. Vacuum harvesters have been designed for crops such as subterranean clover and some tropical pasture plants where seed shedding is impossible to prevent. The crop is usually windrowed and seed may be retrieved from the windrow using a combine with a pick-up. Thereafter the vacuum harvester is used behind a side delivery rake which moves the windrow to one side to expose the soil on which shed seed has fallen. The seed gleaned by suction contains much soil, stones and debris which have to be removed by screen cleaners and aspiration.

Some crops which mature irregularly over a long period can be cut with a combine set to thresh out only the really mature seed. The immature seed is left on the cut plants to mature after which the windrows can be picked up and rethreshed. An alternative which has been used for some grass seed crops is to remove the combine knife and use the reel to knock out the ripe seed at a first pass with the combine; later the crop can be harvested in the normal way with the knife replaced in the combine. In some instances machines which shake the plants and catch the falling seed have been used; these are similar in principle to machines designed to harvest soft fruit.

Forage crops such as white clover (*Trifolium repens* L.) can be very short when ready for harvest. These can sometimes be harvested using a lawn-mower fitted with suitable trays to catch the cut material. Alternatively, a forage harvester can sometimes be used, the cut material being blown into a silage trailer. When this is done it is essential to avoid excess heating in the harvested material as this will impair germination; the whole crop should be dried and the seed separated as quickly as possible. For clover, hullers are sometimes fitted to stationary threshers or the crop may be threshed twice. Combines can usually be set to thresh clover seed satisfactorily.

When conventional machinery is used in an unconventional way to harvest seed crops it is often worth fitting seed-catching trays at strategic points. This applies to binders or mowers used for windrowing and to windrowers. Such trays fitted behind the knives or beneath elevators can often recover a worthwhile quantity of mature seed which might otherwise be deposited on the ground.

For all equipment, cleanliness is essential. All machines used for harvest and the trailers or trucks used to convey the harvested seed from field to store must be carefully cleaned before use. Any bags or other containers used for the seed should also be cleaned; if possible only new bags should be used. Many seed-certification schemes require that new bags only be used.

During harvest, seed placed in bags should be labelled to ensure that its identity is preserved. Other containers which may not be emptied for some time should also be identified. Records of the sequence of operations and weather conditions are worth keeping. A day diary kept during harvest can be invaluable later if any problems arise relating to the germination or cultivar purity of the seed.

THE IMPORTANCE OF SEED MOISTURE CONTENT

In the previous chapter we discussed seed moisture content as one of the factors which decide whether or not seed can be stored safely without loss of germination. When moisture content is too high the seed may heat, and various moulds can grow. Therefore it is absolutely vital to ensure that harvested seed is at a safe moisture content before putting it into store.

A seed crop is almost always cut when the seed moisture content is higher than desirable for safe storage. If the crop is left cut in the field for some time, seed moisture content will be reduced and there will be less green material (leaves, stems and weeds) when the seed is eventually threshed. However, there is no guarantee that seed moisture content will be reduced to a safe level, and when seed is threshed directly in the field by combining, seed moisture content will usually be too high. Only when combining takes place in very dry weather will it be safe not to dry the seed further.

Except in the drier regions where seed crops are grown with irrigation, therefore, a seed grower must plan for some drying and conditioning before seed goes into store. Even in drier areas there may be green material in the harvested seed which could cause damage, and a seed moisture content tester is thus an indispensable tool for every seed grower.

Safe seed moisture contents vary with the crop, but generally 14 per cent or less is considered satisfactory for short-term storage and 10 per cent or less when the seed is to be stored for more than a few weeks. For storage beyond the next growing season, moisture has to be between 5 and 7 per cent, and special storage conditions will be needed.

SEED CONDITIONING AFTER HARVEST

Immediately after threshing seed is usually in a vulnerable state. It will normally contain some debris – broken leaf, stem and chaff – which may have a high moisture content and this aggravates the situation because the seed itself is also often at a moisture content higher than desirable. On no account should seed be left for any length of time before conditioning. Even a few hours in a closed bag or bulk container can cause great damage to germination.

If seed has to be left overnight in bags they should be stood in lines with ample space to allow air circulation; bags should be only half filled and the mouths unfastened and turned back to allow heat and moisture to escape. If at all possible it is preferable to spread the seed thinly on a dry floor; seed depth should not exceed 12 cm and the seed should be turned if it cannot be moved for conditioning.

Conditioning often starts with pre-cleaning to remove some of the debris. For very small quantities hand sieving is possible, while for intermediate quantities a hand or motor-driven winnower can do the job. High-capacity pre-cleaners usually either have a large screen area and do not incorporate aspiration or may be designed for aspiration only. Pre-cleaning is particularly desirable when the seed contains a lot of debris and is to be dried artificially using warm air, because it reduces the amount of material to be dried.

There are many different methods for drying seed. In dry climates it may be sufficient to spread the seed out in the sun on a clean drying floor, turning it frequently. However, the seed must not be allowed to become too hot in direct sunlight. If some rain can be expected, it is better to spread the seed inside a well-ventilated building; drying will take longer and the seed should be lifted for further treatment when it is not raining.

In more difficult climates an in-sack drier or ventilated floor can be used. The in-sack drier consists of a platform above a plenum chamber into which a fan blows warmed air; the platform has a series of gridded holes each the size of a sack. The sacks must not be overfilled and should be turned during drying to avoid moisture accumulation in the upper layers of the sack. A ventilated floor is arranged with ducts to permit air to be blown upwards through seed spread on it. Some systems use underfloor ducts and others lateral ducts placed on a solid floor; the former system may require hessian sheeting over the ducts when small seeds are being dried while the latter will normally require a greater depth of seed so as adequately to cover the ducts.

A further possibility is the ventilated bin. These also are of two types: in one the air is blown upwards from a perforated floor, while in the other the air is distributed from a central perforated tube and dispersed through perforations in the bin walls. In the former type the seed must not be at too great a depth otherwise the upper layers may not be dried before damage occurs. The actual depth will depend on the capacity of the fan and the type of seed (resistance to airflow) and the manufacturer's advice should be sought.

Warm air dries seed by removing moisture from the seed coat. When seed is in a deep layer the bottom part becomes dry, but the air moving upwards becomes progressively moist, so that the top of the layer may actually become moister before it begins to dry since the lower part will be releasing moisture for some time as it is transferred from inside the seed to the seed coat.

Drying must not be too rapid or it may cause the seed coat to harden, trapping moisture inside the seed. Temperature thus has to be carefully controlled. The types of drier so far described are usually controlled so that the air temperature delivered by the fan is no more than 3–5 °C above that of the ambient air, the lower temperature rise being used in drier weather. If ambient air temperature exceeds 40 °C the heater should not be used.

For larger quantities of seed, continuous flow or batch driers can be used. Most types are suitable, although for some of the grass seeds it may be difficult to get them to flow satisfactorily through tower driers and for these a flat bed is preferable. Batch driers can also be used for most types of seed. Airflow may have to be adjusted to avoid blowing light seeds out of the drier.

Temperature of the warm air in a continuous flow or batch drier must be carefully controlled. The higher the initial moisture content of the seed, the lower must be the air temperature and the longer it will take to dry the seed. Seed with a high initial moisture content should preferably be dried twice, otherwise the drier may have to be run so slowly that the seed may be damaged while waiting to go over the drier. Table 3.1 shows some suggested drying temperatures for a range of seeds. Blowing with warm air is followed by cold air; it is very important that seed should be cooled effectively before going into store as otherwise the retained heat will damage germination.

Table 3.1 Suggested maximum warm air temperature for drying seed

Crop	Initial moisture content (%)	Temperature of air reaching seed should not exceed (°C)
Wheat, barley, oats, rye	Over 24	44
	24 and below	49
Brassicas and clovers	18–20	27
	10–17	38
Peas	Over 24	38
	24 and below	43
Chaffy grasses (temperate), e.g. *Lolium* spp., *Dactylis glomerata*	45	38
	40	49
Tropical grasses, e.g. *Chloris gayana*, *Bracharia* spp.	Over 18	32
	10–18	37
	Below 10	43
Soya bean	20	40
Maize (usually dried on the cob)	25–40	35
	Below 25	40

In choosing the type of continuous flow or batch drier to use for seed, the grower must consider the ease with which the equipment can be cleaned down when changing from one kind of seed to another, or when different cultivars of the same crop are to be dried on the same equipment. Flat-bed driers are generally easier to clean than tower or rotary driers because it is possible to inspect the drying area without dismantling the drier. As well as the drier itself, ancillary equipment such as elevators, conveyors and holding bins must be capable of cleaning out between different batches of seed. The time available to do this job is often very limited, but if it is not done thoroughly the quality of the product can easily be spoiled.

SEED CLEANING

When seed has been dried the next step is to clean it. Cleaning has two objectives: to remove seeds of species other than the crop species and inert matter and to select from within the crop seed a finished product graded as to size from which light, discoloured or otherwise unhealthy seed has been removed. Seed cleaning cannot be performed satisfactorily on a rule-of-thumb basis, since each seed-lot presents problems which have to be analysed and solved by using particular machines set in a particular way. The operator of the seed-cleaning equipment can have a greater influence on the standard of cleaning than the equipment itself.

For small quantities of valuable seed hand-picking is as efficient as any machine. Selection may be positive, when the desired seed is picked from the bulk, or negative, when undesired seed or material is removed. To assist hand-picking the seed can be spread on a table, or can be fed on to a moving belt which passes in front of the pickers.

The earliest attempts to clean seed were by winnowing. The seed was thrown into the air over a clean floor in a place where a breeze could be expected. The heavy material fell to earth first, the good seed in the middle and the light seed and chaff was blown the furthest. The principle was subsequently used in machines constructed to blow air through seed falling in a confined space or to suck air from a chamber through which seed is allowed to fall. The latter system, known as 'aspiration' is generally used in most modern machines. Both systems allow heavy and light material to be removed from the crop seed.

The next cleaning process involves placing the seed on a sieve or screen which is then shaken; some material will pass through while other will be retained. The extent to which separations can be made by this method depends on the size of the seed relative to the unwanted material, the size and shape of the perforations in the screen, the speed and distance travelled at each 'shake' and the length of time the seed is subjected to the screening process. Screening by hand-held sieves for small quantities is possible, but not always accurate because the operator may not shake the sieve with a consistent motion or for a consistent time for each batch. Continuous-flow machines which are motor driven are therefore generally preferred. The screens may be in a flat bed, and these machines usually incorporate at least

Table 3.2 Examples of perforation sizes in screens (mm)

Crop	Removal of oversize material	Removal of undersize material
Cereals	20 × 4.5 or 20 × 4.0 slot 6.0 or 5.5 round hole	20 × 1.5 to 20 × 2.6 slot
Peas	9.5 round hole	25 × 5.0 slot
Rape	3.5 round hole	20 × 1.2 slot 1.2 round hole

Source: Toveys of Cirencester Ltd, *Seed Cleaning Recommendations*.

two screens – an upper one which retains large material and a lower one which retains the good seed and allows small material through. Machines with more screens are also available. An alternative is the rotary screen which is cylindrical and arranged in sections with different shapes and sizes of holes; the seed is fed into the rotating screen and as it moves along inside various fractions are removed. Screens require cleaning during operation, and this is usually achieved by arranging for them to be tapped by hammers or brushed during operation to prevent clogging.

Screen perforations can be of various shapes and sizes, but the most usual are round holes or slots; some examples of sizes are given in Table 3.2 but individual machines or particular cleaning problems may require different sizes.

Most modern seed-cleaning machines combine aspiration and screen cleaning in one machine, as illustrated in Fig. 3.3, and separate fractions on

Fig. 3.3 Air/screen cleaner (Photograph: Toveys of Cirencester Ltd)

the basis of weight (aspiration) and size/shape (screens). Machines are available with various capacities and may be hand operated or, for larger capacities, motor driven. Actual capacity of an individual machine will depend upon the complexity of the cleaning problem and the kind of seed (larger seeds can normally be cleaned faster than small ones).

For some seed-cleaning problems, separation on the basis of length is required and there are two machines which have been designed for this. The most widely used is the 'indented cylinder'. This consists of a hollow

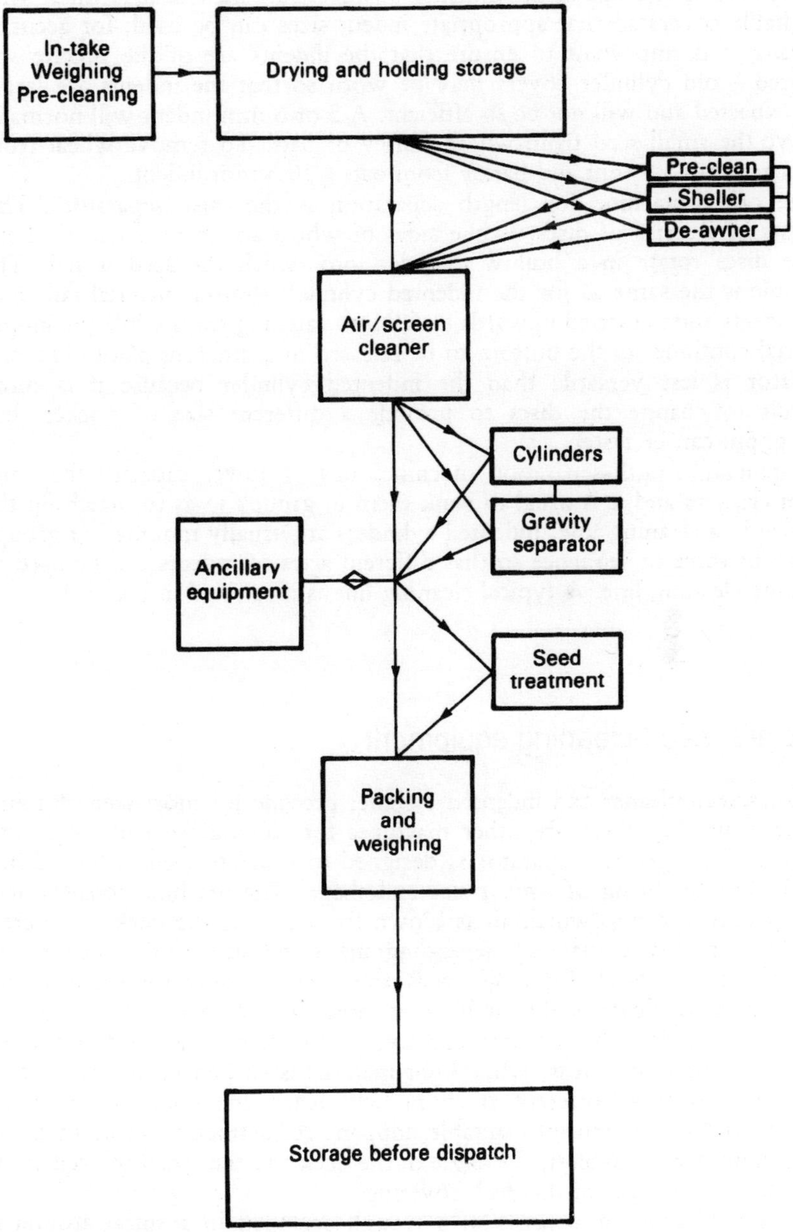

Fig. 3.4 Seed-cleaning line

cylinder of sheet steel, the inside of which is covered with indents of a precise size. Through the middle of the cylinder runs a trough in the bottom of which is an auger. The seed is fed into the bottom of the cylinder which is rotated: shorter material lodges in the indents and is carried upwards by the rotation to fall into the centre trough and be removed by the auger; the longer material remains in the bottom of the cylinder and passes right through. The efficiency of the machine depends upon the rate at which seed is fed through, the height of the trough above the cylinder floor (which is adjustable) and the size of the indents. Cylinders are now usually made with detachable covers so that appropriate indent sizes can be used; for accurate working it is important to ensure that the indents are of the precise size required – old cylinder covers may be worn so that the indents are larger than expected and will not be so efficient. A 5 or 6 mm indent will normally remove the small seed from wheat, barley or oats. To remove wheat from barley requires a 9 mm and barley from oats a 10.5 mm indent.

The other machine for length separation is the 'disc separator'. This consists of a series of discs, in the sides of which are many small pockets. These discs rotate in a hollow cylinder into which the seed is fed. The principle is the same as for the indented cylinder: shorter material falls into the pockets and is carried upwards to fall into catching trays while the longer material continues in the bottom to be released in a different place. The disc separator is less versatile than the indented cylinder because it is more difficult to change the discs to provide a different size of pocket, but throughput can be faster.

In general, length-separation machines have a lower capacity than air/screen cleaners and it is usual to bank them in groups so as to match up the capacity in a cleaning line. Indented cylinders are usually mounted in groups of two or three in sequence so that different sizes of indents can be used in the same cleaning line. A typical cleaning line is illustrated in Fig. 3.4.

Specialist seed-cleaning equipment

The air/screen cleaner and indented cylinder provide for most seed-cleaning requirements, but there are other machines for solving particular cleaning problems. The gravity separator is designed to separate seeds which differ only in weight, being of similar size and shape. The machine consists of a grading deck through which air is blown from below; the deck is covered with an appropriate material depending upon the size of the seed to be cleaned – coarse mesh for large seeds and a cloth cover for small seeds. Lighter material floats in the air-blast, whereas heavier material remains on the deck. The deck is oscillated and can be tilted, so causing the heavier material to move upwards, while light material is suspended in the air and travels downwards; different fractions will reach the outlet in different positions and can be fed into suitable hoppers. Adjustments can be made to the strength of the air-blast, the angle of the deck and the speed of oscillation in addition to changes of the deck covering.

The spiral separator consists of a trough arranged in a spiral around a vertical axis; when seed is allowed to run down the trough, round seed moves faster to the outside than flat seed and can be directed into separate spouts by divisions in the bottom of the trough.

Seed-coat texture can be used to separate some seeds. A rough seed coat or a seed with hooks, barbs or hairs will attach to certain cloth coverings. By feeding seed on to a moving belt covered with suitable cloth which is inclined upwards, some seeds can be carried upwards and removed by brushes at the top roller, whereas the smooth coated seed travels down. Some machines for extracting dodder from clover have two rollers together which rotate in opposite directions so that the seed is in a trough between them and given a greater opportunity to adhere to the cloth.

For the magnetic separator the seed is mixed with a small amount of liquid containing iron filings; this adheres to some seeds but not to others, depending on the seed-coat characteristics e.g. to separate the weed *Galium aparine* (cleavers) from brassica seed. Seed so treated is then passed over a strongly magnetised roller so that those with iron filings cling to the roller and can be scraped off and separated from the main stream.

Differences in electrostatic characteristics can be used to make some separations; when allowed to fall through an electrical field some seeds are attracted to the positive and so are diverted from the main stream to be caught in a different spout.

Marked differences in colour can also be used to make some separations, for example for removing discoloured (brown) peas from an otherwise pale green seed-lot. The seed is allowed to fall in a single stream through a chamber containing photoelectric cells which scan the falling seeds from each side. When a discoloured seed is detected a jet of air is activated to direct that seed from the main stream.

The needle drum separator consists of a drum inside of which are a number of protruding needles. As seed is passed through the rotating drum seeds which have holes in them (e.g. peas damaged by weevils) are carried up on the needles and removed by a rotary brush to fall into a central trough.

De-awners are used to remove excess awns, tips or loose glumes by beating or brushing the seed against a prepared surface. They can improve the appearance of some samples (e.g barley) and make it easier for seed to flow from the hopper of a seed drill, so making it possible to achieve a more uniform distribution of seed in the field at sowing time.

Ancillary equipment and operation

A seed-cleaning plant requires careful planning to achieve the most efficient and economical working. The separate machines are linked together by conveyors and elevators, and at appropriate points feed controllers are installed to ensure that seed enters each machine in a steady, controlled stream. This ensures that no machine is under- or overloaded, either of which can cause loss of cleaning efficiency. All machines and ancillary equipment should be easily accessible. When changing from one kind of seed to another or from one cultivar to another the cleaning line will have to be cleaned out thoroughly so as to avoid unwanted mixing. When changing from one cultivar to another of the same crop it is advisable to clean a lot from a different crop in between so that any mixture can be easily seen, and also to discard a small quantity of the first seed through the equipment.

Seed cleaning should be closely associated with quality control. A seed-testing laboratory can analyse the problems for each seed-lot and monitor the success of the cleaning process during operation.

An important aspect of quality control is to keep good records. A day-book should be used to record all work done each day, and particularly the sequence of events on each machine. Such records are invaluable subsequently when it may be necessary to check the possibility that some mistakes were made during the seed-cleaning operation.

SEED SAMPLING

The control of seed quality during seed cleaning depends upon an assessment of the state of each seed-lot before and after cleaning. These assessments are made on samples drawn from a seed-lot. A sample is defined in the Concise Oxford Dictionary as 'small separated part of something illustrating the qualities of the mass etc. it is taken from'. The important word in the definition is 'illustrating'. If the illustration is to be accurate, we must ensure that the sample truly represents the whole. However, a seed-lot is rarely uniform within itself. Seed crops usually contain areas which differ from one another – for example, parts of the field infested with weeds or infected by disease – and during harvest the seed is not thoroughly mixed; seed being moved in bags or in bulk on conveyors tends to segregate as the various movements affect large and small seeds differently.

Sampling, therefore, has to be done in such a way that the effects of irregularities within the seed-lot are minimised, and to do this certain procedures should be followed:

1. Ensure that a seed-lot is as uniform as possible before sampling. If it appears to be unduly heterogeneous, mix it throughly first.
2. Do not take a sample from too large a lot. Table 3.3(a) shows suitable sizes.
3. Take subsamples from all parts of the lot. Table 3.3(b) shows suitable numbers of subsamples for various lot sizes.
4. Use a suitable sampling tool (stick or sleeve-type trier) and ensure that it takes samples from all parts – top, middle, bottom, centre and sides – of containers. Sampling by hand is less satisfactory except for some chaffy grasses.
5. Mix the subsamples thoroughly to make a representative sample.

In some cleaning plants mechanical samplers are provided in the seed-conveying system. These normally direct the seed flow to a collecting container at regular intervals; it is important that the whole flow is diverted because samples taken from the side or centre of a spout may be biased towards a particular size, shape or weight of seed according to how the seed flows in the spout.

Table 3.3 Sampling: maximum lot sizes and numbers of subsamples to be drawn

(a) Maximum lot size to be represented by one sample

Seeds less than the size of wheat	10 000 kg
Seeds the size of wheat or larger	20 000 kg
Maize only	40 000 kg

(b) Intensity of sampling

Number of bags	Number of subsamples	Weight of lot (kg)	Number of subsamples
Up to 5	At least 5	Up to 500	5
6–30	5 or 1 in 3 whichever greater	501–3000	1 in 300 kg, but at least 5
31–400	10 or 1 in 5 whichever greater	3001–20 000	1 in 500 kg, but at least 10
Over 400	80 or 1 in 7 whichever greater	Over 20 000	1 in 700 kg, but at least 40

Source: ISTA (1985).

SEED TREATMENT

Seed may be treated before sowing for several different purposes:

- Seed disinfection is intended to combat seed-borne pests and diseases.
- Seed protection is intended to protect the seed against pests and diseases which may be present in the soil or be airborne when the seedlings emerge.
- Seed coating is a way of adjusting the size of some irregularly shaped seeds to make it easier to sow them, and may also be a means of providing disinfection or protection and some plant nutrients.

The chemicals or other treatments used for disinfection or prophylaxis are often toxic to the seed and to humans and animals. Treatment must therefore be done with great care and precision. Usually seed is treated just before it is needed for sowing so that it is not stored for long periods after treatment: this minimises the risk of damage to the germination capacity and reduces the risks that excess treated seed may be used as food for humans or animals. Treated seed should always be clearly labelled to show what has been applied and to warn people of any dangers; a graphic picture showing the danger should be included on the label.

Seed-borne pests and diseases may be carried within some seeds (for example, the mycelium of loose smut, *Ustilago tritici*, in wheat) or on the seeds (for example spores of bunt, *Tilletia* spp., of wheat); or may accompany the seed as free-living organisms or in debris. The most difficult to combat are the first, since the treatment must penetrate the seed; in the past, heat treatment has been used as in the hot-water soak for loose smut, but today systemic fungicides which are translocated to the infected site have been developed. External infections can be attacked directly; until recently the most commonly used treatment against fungal infections was with organo-mercurial dressings, but because of the toxicity of mercury compounds these are largely being replaced by other formulations such as the dithiocarbamates. Some free-living organisms (e.g. Sclerotia) and debris can be removed during seed cleaning, but other infections may require chemical treatment or fumigation (e.g. nematodes).

Protecting seed and seedlings against possible attack in the soil or against the aerial parts after emergence has been successful in some cases using systemic insecticides or fungicides. Carbofuran applied to sorghum seed, for example, has given some protection against shootfly (*Atheriyona soccata*). Benomyl has given control of powdery mildew (*Erysiphe graminis*) in barley for some time after seedling emergence.

Seed coating or pelleting is little used in the agricultural crops generally, but is now widely adopted for fodder and sugar-beet. The advantage is that irregularly shaped seed can be converted to a regular shape by the process, so making it possible to sow them at a regular spacing. Special equipment is required, and the process is normally undertaken by specialist companies.

The application of treatments to seed requires some equipment. The chemicals may be in the form of powders, slurries or liquids, and the problem is to apply exactly the right dose to each seed. Hand-mixing of seed by turning a heap to which the chemical has been added is not generally very efficient. An alternative is the drum mixer, which consists of a clean oil drum

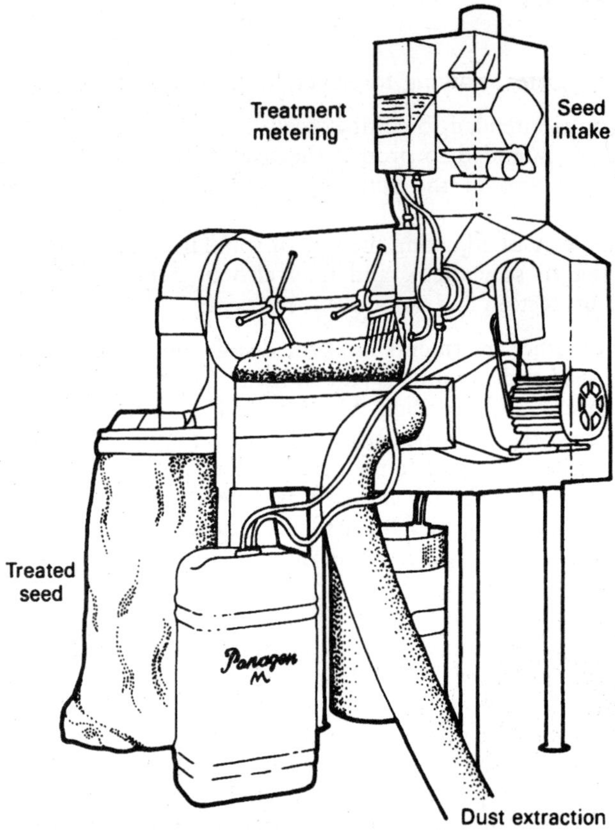

Fig. 3.5 Seed-treatment equipment (Photograph: Shell Chemicals UK Ltd)

mounted eccentrically into which seed and chemical are placed; the drum is then rotated for some minutes. Concrete mixers are also sometimes used. Equipment designed for treating seed is available for placing in the seed-cleaning line. (Fig. 3.5). The mixing is usually achieved either by an auger, or by a rotary action: it is important that the seed receives the correct dose, so that both seed and chemical have to be metered into the mixing chamber. Some machines work on the batch system, where a weighed quantity of seed is released into the mixing chamber together with the appropriate amount of chemicals; others are continuous flow with both seed and chemical being metered continuously.

Powders are the most difficult to handle and to mix, although they are convenient to transport. Slurries and liquids are easier to mix, but may cause damage to the seed if handled incorrectly. Most seed-treatment machines are now designed to handle all forms of chemicals.

As we have noted earlier, seed-treatment chemicals are usually toxic to humans and animals and must be handled with great care. Operators must wear appropriate protective clothing and face-masks and should be instructed in the hazards of the task, and made aware of the precautions necessary and remedial action to take in case of accident. Seed-treatment equipment should be installed with suitable precautions such as dust extraction to protect the operators. In the event that fumigation is needed it is usually necessary to call in specialist operators as the fumigants used are highly toxic and require special fumigation chambers (e.g. methyl bromide for nematodes in clover seed).

Seed inoculation

In some situations, seed of leguminous crops requires inoculation with *Rhizobia* spp. to ensure satisfactory nitrogen fixation. Many of the chemical seed treatments are incompatible with inoculation since they are toxic to the *Rhizobia* spp., and care should be exercised in the selection of suitable treatments in such cases. Inoculum is best added to the seed just before sowing and is therefore normally done on the farm.

4 SEED MARKETING AND DISTRIBUTION

In the preceding chapters we have discussed the principles of seed production. These principles apply when seed is produced by farmers for their own use and when it is produced for marketing on a wider scale. However, there is little point in producing seed for marketing unless we can see clearly where and when it will be used (Fig. 4.1). Seed is a perishable commodity, expensive to produce, store and transport, and therefore production must be geared to realistic marketing and distribution targets. In assessing market needs we first have to answer the questions:

- How much, where and when?
- To what standards of quality?
- At what price?

When we have the answers to these questions we can plan both the scale and the timing of production to satisfy the expected demand.

Fig. 4.1 Seed is required to produce crops for a particular market

HOW MUCH, WHERE AND WHEN?

It is usually relatively easy to calculate the total seed requirement for a particular crop in a specific country. Statistics of crop areas are generally available and local seed rates are known. However, total seed requirement is of little use by itself. We now have to estimate what proportion of that total requirement can be supplied through the market as opposed to farmers using seed they have saved themselves, and of this we must further estimate how much can be supplied by the cultivar or cultivars which we are considering. These estimates require market research, involving enquiry of the farmers who are expected to acquire the seed. We need to know what motivates the farmers to buy seed, what their purchasing power is likely to be and how many potential customers there are in a particular area.

Motivation to buy seed will depend primarily upon convincing the farmer that such a purchase is advantageous. The most effective way to do this is to produce convincing trial results within the area (preferably where the trials can be shown to the potential customers) showing the advantages of a particular cultivar in terms of better yield, quality of produce or other cultural advantage. Practical on-farm demonstrations are recognised as having the greatest value, but additionally promotion can be organised through the conventional methods of catalogues, press articles, radio or television and so on. These activities announce a forthcoming production and from them there should be feedback of information to permit a realistic assessment of the quantity of seed which it will be possible to dispose of and where the market outlets ought to be located.

The timing of the market operation will depend upon the multiplication possibilities. As we have discussed in earlier chapters seed has to be multiplied through several generations before sufficient is obtained for marketing. To plan production to meet a predetermined quantity we have to know the quantity available from the cultivar maintenance programme and the multiplication rate. From these data we can estimate how many growing seasons are needed to reach the required quantity for market, making due allowance for possible failures or other losses during multiplication.

With all this information it is possible to plan production by translating the target quantities required each year into areas of seed crop required. Normally, initial multiplication steps will be taken while market research is still proceeding. The quantities will be small enough so that over-production is avoided and it is important that the interval between advance publicity – inevitable in estimating market potential – and actual availability of seed on the market should be as short as possible. Once customers have been made aware of a potentially valuable new cultivar they will wish to try it as soon as possible and may lose interest if kept waiting too long.

Production should be planned to provide continuity. Once a demand has been stimulated it must be satisfied each growing season for as long as that particular cultivar remains popular. Each year, therefore, we have to estimate demand at the next growing season and plan production to satisfy that demand. As new and improved cultivars become available they have to be integrated into the production programme, while obsolete cultivars will be dropped. Example 4.1 shows a hypothetical calculation for a new wheat cultivar.

Example 4.1 Production programme calculations.

Crop: wheat. Estimated market demand for seed: 1000 tonnes
Average seed rate: 100 kg/ha
Average yield after harvest/cleaning loss: 3000 kg/ha
Area required to produce 1000 tonnes: 333 ha
Seed required to sow this area: 33.3 tonnes
Area required to produce 33.3 tonnes: 11.1 ha
Seed required to sow this area: 1.1 tonnes (this quantity should be produced from the maintenance programme for the cultivar)

To allow for continuity we can construct a table as follows:

Year	1	2	3	4	5	6	7
Area		11.1	333				
Quantity	1.1	33.3	1000				
Area			11.1	333			
Quantity		1.1	33.3	1000			
Area				11.1	333		
Quantity			1.1	33.3	1000		
Area					11.1	333	
Quantity				1.1	33.3	1000	
Area						11.1	333
Quantity					1.1	33.3	etc.

Because the multiplication rate is × 30 the quantities remain quite small up to the second year; it may therefore be justified (depending on costs and prices) to proceed this far while market research is continuing, i.e. up to year 2. Similarly, adjustments can be made to the programme in subsequent years to provide more or less seed for the market depending upon the actual demand in the previous season. For self-fertilised crops such as wheat it is possible to continue multiplication for longer than shown in the table by retaining part of the 1000 tonnes for further multiplication, but this is not possible when generations are limited in cross-fertilised crops.

Special needs of hybrid cultivars

When producing a hybrid cultivar it is necessary to follow a particular formula. In so doing the production programme has to be planned in such a way that in the year in which hybrid seed is produced to meet market needs sufficient quantities of each of the parents are available. This will involve making separate calculations for the male and for the female parents. Usually less seed of the male parent will be required than of the female, although there are exceptions. Planting ratios are discussed in the chapters dealing with individual crops where appropriate.

TO WHAT STANDARDS OF QUALITY?

In deciding upon standards of quality we have to take into account first the needs of the customer and second the difficulty and cost of achieving a particular standard. Quality standards should have practical value for the customer; standards which are too high increase costs needlessly, while those which are too low serve no useful purpose. There is thus a conflict between the desire to make claims for high quality as an aid to sales, and a compromise situation where quality claims are realistic in relation to actual needs. Quality standards can be established for the following:

1. Cultivar authenticity and purity.
2. Analytical purity.
3. Weed seed content (based on seed count per unit weight).
4. Germination capacity.
5. Any other quality characteristic important in the area where the seed is to be marketed (for example, moisture content, infection with seed-borne disease).

Of these various quality characteristics the most difficult is germination. Once seed has been sealed in containers it is not possible for changes to occur in cultivar purity, analytical purity or weed seed content. However, germination can deteriorate in store and a test result can therefore only be valid for a limited period. In practical terms this period can vary from about a year in more temperate or very dry areas to much shorter periods in the more humid areas of the tropics; obviously the conditions under which the seed is stored will affect significantly the period for which a germination test result will remain valid.

Seed health standards are also difficult to establish. Usually the number of seeds which can be examined for a particular disease is very limited so that it is difficult to relate the test result to what is actually present in the seed-lot – a nil result does not necessarily mean that there is no infection present. Standards thus have to represent compromise between what is desirable and what is practical.

AT WHAT PRICE?

Traditionally, seed has been relatively cheap considering the benefits which good-quality seed of an improved cultivar can confer. Usually seed cost constitutes no more than 10 per cent of the cost of growing a crop. Prices are subject to severe competition and, particularly for the self-fertilising crops, the farmer always has the option of saving seed to be used next season.

Price, therefore, has to be calculated to cover direct and indirect costs of production plus a reasonable profit, but must also have regard to what the customer will pay. The customer must perceive some benefit at the price being asked.

For those crops where seed is also the crop produced for food or other purposes (cereals, oil-seeds, dry seeded legumes), price calculations usually start by taking the market price of the product as the basic production cost. To this is added a seed-grower's premium, to cover the extra expense such as cleaning equipment before sowing and harvest, safeguarding isolation, roguing and so on. Next will be added the seed-drying, cleaning, treatment and packaging charges, together with storage and transport, plus an allowance for seed loss during cleaning. Finally, there have to be added the costs of quality tests, advertising or other promotion costs and reasonable profit. Cost calculations should include allowance for depreciation on plant and buildings, interest on capital invested and so on. There may also in some countries be a plant breeders' royalty where plant breeders' rights legislation is in force.

This will normally give a price which is at least 50 per cent above the selling price of the product, and can be as high as 400 per cent above when the cultivar shows very marked advantages and seed of it is to be sold for the first time.

For other crops where seed is not usually the product for which the crop is grown (forage and lawn grasses, forage legumes, forage brassicas, beet) there is no overall market price on which to base calculations. The seed-grower's remuneration has to be based on an estimate of costs of seed production, but the other cost elements in making up the price will be the same as for the previous group of crops.

SEED DISTRIBUTION

In the marketing process we have now decided upon the quantity of seed to be sold, the area where the sale will take place, the growing season and year for the sale; we have also decided the quality to be aimed at and the price at which the offer for sale will be made. We now have to decide upon the logistics of how the seed is to be distributed. The object should be to keep the seed in central storage until the last possible moment because storage on farms is likely to be less suitable and seed could deteriorate if not sown soon after delivery. This may not be wholly true where farms are large and well organised, and in these situations it may be advantageous to start deliveries earlier. Where a large central store is expected to serve a very large area, it may be necessary to set up a network of subcentres where short-term storage is available.

Transport should be as quick as possible. Seed is vulnerable during transit, particularly in difficult climates such as the humid tropics. Supply also needs to be sufficiently flexible so that surplus seed in one area can be moved quickly to supply a need in another.

Distribution has to be arranged so as to deliver the right quantity of seed of the right quality in time for sowing when conditions are at their best. This makes it essential to give careful thought to the delivery system so that it functions smoothly in the limited period available for moving the seed. Because the business is seasonal, it is not usually economic to have a transport system solely for seed. Either the distribution system must be used also for the supply of other goods – fertiliser or feeding-stuffs for example – or contractors will have to be used.

SELLING SEED TO FARMERS

Seed does not sell itself. For many crops the farmer has the option of saving seed for own use. We have to make the effort to offer convincing evidence of the advantages of making a purchase in terms of better yield, better quality of product and better financial return from the crops which the seed will produce.

In general terms the seed trade may be divided into the following categories:

1. The plant breeder who produces the new cultivars and limited quantities of seed from a cultivar maintenance programme.
2. The wholesaler who takes the seed from the plant breeder and multiplies it (usually by contracting with farmer seed growers) to a marketable quantity.
3. The retailer who buys seed from the wholesaler and sells it to the farmers.
4. The brokers who sometimes act as intermediaries between the wholesalers and the retailers.

It is, of course, possible for two or more of these functions to be combined in a single organisation or company, and some larger commercial enterprises combine all four. Many wholesalers are specialised, dealing only in a limited number of crops; retailers usually cover a wide range of crops and may also supply farmers' other requirements. The trading functions and seed distribution may be performed by a government agency, by farmers' co-operatives or by private persons or commercial companies.

To sell seed there are various ways in which the farmer/customer can be approached:

- Demonstrations, talks to groups of farmers, trade stands at shows.
- Advertising in press, on radio and television and through posters.
- Individual approach by salespersons.

In many developing countries there are various government subsidies to encourage the use of good-quality seed. Often these take the form of cheap credit arrangements, so that the farmer is not required to pay for the seed until the crop grown from it has been harvested; alternatively, there may be barter arrangements whereby the farmer can exchange grain for an equivalent amount of good seed. In a market economy, however, the seed has to be sold at a price which will give a reasonable return to those concerned in its production and marketing. In all cases it is important that any government extension or advisory services should be fully briefed on the benefits of using good seed, and particularly of the proven attributes of the cultivars on offer.

The efficiency of practical demonstration has already been stressed. Sometimes this can be combined with cultivar trials conducted in the area where seed sales are to be expected. Alternatively, a demonstration may be laid out in a farmer's field as a means of convincing others of the practicality of the new cultivars. In some developing countries a system of 'mini-kit' demonstration has been evolved: enough seed and other inputs for a plot (perhaps half a hectare) are supplied in a package with instructions to leading farmers in each area where the cultivar is to be sold and in this way local people are encouraged to watch the progress of the crop.

Advertising should be factual, and wherever possible should rely upon trials evidence to speak for the cultivars. As part of advertising it is important that seed be delivered in attractive packages or bags with a clean, smart appearance.

To make a sale, however, each farmer has to be approached individually or be persuaded to approach a salesperson of the organisation offering the seed. These persons need to know the details of what they are selling and must be prepared to discuss the relative merits of different cultivars on offer.

The success of an organisation selling to farmers depends mainly on the reputation which that organisation is able to build up over the years for supplying good-quality seed of reliable cultivars. Consistency of performance from year to year is generally more highly regarded by farmers than an outstanding performance in one year followed by partial failure or disaster in another. This is summed up in the following extract from the memorandum of association of the Seed Trade Association of the United Kingdom (Central Office of Information, 1961): 'To improve and perfect a standard of business integrity which shall include purity of stocks, honesty of representation, carefulness of obligations, and promptness and efficiency in execution.'

PART 2 SEED GROWING OF PARTICULAR CROPS

5 CEREALS

SELF-FERTILISED CEREALS

Triticum aestivum L. emend Fiori et Paol. Ploidy : hexaploid. Common name: wheat; bread wheat

Classification of cultivars

Seasonal type	Winter; Alternative; Spring
Intended use	Animal feed; Biscuit flour; Bread flour

Main distinguishing characteristics of cultivars

Plant growth habit	Erect; Prostrate
Flag leaf attitude	Rectilinear; Recurved
Time of ear emergence	Early; Late
Glaucosity of flag leaf sheath	Weak; Strong
Glaucosity of flag leaf blade	Weak; Strong
Glaucosity of ear:	Weak: Strong
Glaucosity of neck of culm	Weak; Strong
Plant height (stem and ear)	Short; Long
Straw section half-way between base of ear and next node	Pith thin; Pith thick
Ear colour at maturity	White; Coloured
Ear shape	Tapering; Parallel; Fusiform; Clavate
Ear density	Lax; Dense
Awns or scurs	Both absent; Scurs present; Awns present
Distribution of awns or scurs	Tip only; Upper half; Whole length
Scurs at tip of ear	Short; Long
Awns at tip of ear	Short; Long
Extent of internal hairs on lower glume	Weak; Strong
Grain colour	White; Red
Grain coloration with phenol	Light; Dark

The seed crop

Isolation

Wheat is normally self-fertilised and an isolation distance of 2 m is sufficient when there is no physical barrier between crops.

Previous cropping

It is desirable to allow an interval of 2 years free from cereals before growing a seed crop of wheat except when the same cultivar is grown and the seed used to sow the crops can be authenticated.

Difficult weeds

The following weed species are listed and given appropriate standards in the seed-certification schemes of the countries listed:

European Economic Community (EEC) *Raphanus raphanistrum*; *Agrostemma githago*; *Avena fatua*; *A. sterilis*; *A. ludoviciana*; *Lolium temulentum*
Pakistan *Convolvulus arvensis*; *Carthamus oxyacantha*; *Asphodelus* spp.; *Avena fatua*
Brazil *Brassica* spp.; *Raphanus raphanistrum*
Canada *Cuscuta* spp.; *Convolvulus arvensis*; *Helogeton glomeratus*; *Cardaria* spp.; *Solanum carolinense*; *Euphorbia esula*; *Centaurea repens*; *Senecio jacobaea*.

Hormone weed-killers should be used with care, since they can cause distortion of the wheat ears, making off-type recognition more difficult.

Crop layout

There are no special requirements and a seed crop can be handled as any other wheat crop. Tramlines, or gaps left at regular intervals during drilling will help with crop inspection and roguing by providing room for walking through the crop. A low seed rate and wider spacing between the rows can be used when it is desired to obtain the highest possible multiplication rate, but the yield per unit area may be less: seed rates as low as 30 kg/ha have been used successfully in fertile conditions. In many areas grass weeds can be particularly troublesome and it is useful practice to cultivate round the edge of the crop as most will spread from hedgerows.

Crop management

Seed crops do not require any different management from normal, except the attention to the details as outlined in the previous chapters. Fertilisers and straw-shortening chemicals should be used with care. Nitrogen should not be applied to excess as it is necessary to avoid lodging. Pest and disease control should be practised so as to ensure a healthy crop with well-filled seed.

Seed-crop inspection

The best time to assess cultivar purity is after ear-emergence when the seed has started to fill (growth stage 75–85). Later inspections when glume and seed colour can be observed are also sometimes desirable.

Harvesting

Seed crops are harvested at the same stage of ripeness as other crops. The seed should be hard and difficult to mark with a thumb-nail; the straw will have lost its green colour. Combining at moisture content above 20 per cent is not recommended as the seed can suffer mechanical injury; very dry seed below 8 per cent moisture is also easily damaged.

Seed after harvest

Drying

Seed can be stored at 14 per cent moisture when it is to be used in the next season; in northern latitudes when sowing of winter wheat takes place within a few weeks after harvest, 16 per cent is possible. For longer storage moisture content should be reduced below 10 per cent; when it is desired to keep samples for many years they should be dried below 8 per cent before being placed in moisture-proof packages. The temperature of the drying air should not exceed 44 °C when initial moisture content is high; this can be increased to 49 °C when the seed is dried.

Seed cleaning

An efficient air/screen cleaner is essential equipment for the seed producer. Next in importance are the indented cylinder and the gravity separator. One difficulty with winter wheat in northern latitudes is the very short time available between harvest and sowing time; seed-cleaning capacity therefore has to be sufficient to deal with the peak loading for only a few weeks in the year.

Seed treatment

Chemicals used for seed treatment must be handled with extreme care and the manufacturers' instructions must always be followed. Seed above 16 per cent moisture should not be treated. Seed which has been treated must be labelled to show the treatment which has been applied. It usually has only a limited life in store and it is best to treat only as much seed as is likely to be required. The main seed-borne diseases of wheat and current treatments are as follows:

Disease	Treatment
Brown foot rot and ear blight *Fusarium* spp.	Organo-mercury, guazatine, benomyl, thiophanate-methyl, triadimenol/fuberidazole/imazalil
Bunt or stinking smut *Tilletia caries*	Organo-mercury, Benomyl–thiram, carboxin–thiram, guazatine, triadimenol/fuberidazole/imazalil
Glume blotch *Septoria nodorum*	Guazatine, triadimenol/fuberidazole/imazalil
Loose smut *Ustilago nuda*	Carboxin, benomyl, triadimenol/fuberidazole/imazalil

Loose smut may also be treated by a hot-water soak, but this is cumbersome. The seed should be soaked at 32 °C for 4 hours, and then immersed in water held at a constant 54 °C for 10 minutes; after treatment the seed must be cooled and dried immediately. There are also some pests which, although not seed borne, can be combated by seed treatment as follows:

Wheat bulb fly *Leptohylemgia coarctata*	For early sowing: Carbophenothion, chlorfenvinphos For later sowing: gamma-HCH
Wire-worm *Agriotes* spp.	gamma-HCH (note that this is a different formulation from that used for wheat bulb fly)

When slugs or snails are excessive in a field seed may be mixed with methiocarb pellets, but cannot be kept for more than 3 months after mixing.

Although seed treatments can be very effective, many of them, except the organo-mercurials, are also expensive; the seed grower should therefore endeavour to grow a crop free from seed-borne diseases by starting with healthy seed and practising good field hygiene. The organo-mercurials are cheaper and effective, but there are now objections to the widespread use of mercury and it has been banned or restricted in many countries.

Hybrid cultivars

There has been some interest in hybrid cultivars in North America based on cytoplasmic male sterile lines. Some difficulties have been experienced in large-scale seed production. In recent years there has been interest in hybrid wheat cultivars in the USA and Europe based on the use of chemicals to emasculate the desired female parent. Although early work has shown promise, there have been difficulties in achieving satisfactory seed production on a wider scale, partly because of poor availability of pollen at the correct time to achieve fertilisation. The production of hybrid cultivars of wheat can pose particular problems of quality control: it is desirable to find some way of determining how efficient emasculation has been. Tests are being made of a system which would involve covering certain small areas in the female in polythene 'tents', or bagging individual ears; by this mean it may be possible to estimate the amount of self-fertilisation in the protected ears.

Triticum durum Desf. Ploidy : tetraploid Common name: Durum wheat

Classification of cultivars

Seasonal type — Winter; Alternative; Spring

Main distinguishing characteristics of cultivars

In addition to the characteristics listed under *T. aestivum*, the following may also be used:

Hairiness of middle third section of rachis	Weak; Strong
Lower glume shape (mid-third of ear)	Round; Ovoid; Elongated
Lower glume keel spicules (mid-third of ear)	Few; Numerous
Colour of awns	White; Red; Brown/Black
Anthocyanin coloration of awns	Weak; Strong
Divergence of awns	Parallel; Divergent

Grain colour in *T. durum* may also be yellowish or brown/black in addition to white or red listed under *T. aestivum*.

The seed crop

The points listed under *T. aestivum* apply to *T. durum*. Additionally it is necessary to take particular care at harvest because *T. durum* is very prone to sprouting in the ear, which can cause loss of germination after harvest. In difficult climates it is advisable to harvest the seed before the moisture content falls below 18 per cent since subsequent wetting after maturation has proceeded further may cause germination to start.

Triticum spelta L. Ploidy : hexaploid. Common name: spelt

This species is not widely grown, but there are a few crops in Europe, particularly in difficult or mountainous areas. The seed remains enclosed in the glumes after threshing.

Oryza sativa L. Ploidy : diploid. Common name: rice; paddy

Subspecies: *indica* – short day; grown mainly in warm humid tropics; *japonica*: some short-day cultivars, mostly day neutral; grown outside tropics; *javonica*: day neutral; grown in equatorial climates of Java and Indonesia.

There may also be hybrids between these subspecies, although there are difficulties in establishing fertile hybrid cultivars in some combinations.

Classification of cultivars

Photoperiod sensitivity	Insensitive; Intermediate; Sensitive
Water regime	Upland; Shallow; Intermediate; Deep (some cultivars may be adapted to more than one water regime)
Plant type	Semi-dwarf; Tall; Floating
Starchy endosperm	Not glutinous; Intermediate; Glutinous; Mixed

Main distinguishing characteristics of cultivars

Pubescence of blade of penultimate leaf	Weak; Strong
Anthocyanin coloration of auricle	Absent; Present
Time of 50 per cent heading	Early; Late
Stem length (excluding panicle)	Short; Long
Length of decorticated grain	Short; Long

The seed crop

Isolation

Rice is normally self-fertilised. For upland rice a distance of 2 m is required when there is no physical barrier at the edge of the crop. For rice grown in shallow, intermediate or deep water (swamp or floating rice) seed crops will normally occupy a whole enclosure or bund.

Previous cropping

Rice is often grown continuously on the same land; difficulties for seed growing will occur when it is desired to change from one cultivar to another, or when there is a build-up of the weed red rice (*O. rufipogon*). When cropping is continuous the seed grower should ensure that the same cultivar is grown each time and that only authenticated seed is used. When changing cultivars, or when weed rice becomes too prevalent, an interval of 2 years free of rice is needed to guard against volunteer plants.

Difficult weeds

The most difficult weed is *O. rufipogon* which can cross-fertilise with *O. sativa*. According to Purseglove (1985), wild red rice seed is virtually impossible to remove from the crop seed and causes difficulties during milling. Control is also virtually impossible in a rice crop: hand roguing can only be undertaken after grain formation. The seed grower thus has a duty to maintain land for seed crops free from wild rice and to do everything possible to avoid contamination in the seed crop. Seed in small quantities – early stages of cultivar maintenance – can be hand-picked; Purseglove (1985) suggests soaking the seeds in water to enhance the red colour before hand-picking. He also suggests the following methods as giving a partial control:

- never allow a ratoon or dropped seed crop to be taken;
- plough early after harvest to encourage dropped seed to germinate, then graze;
- plough deep to discourage germination of red rice seed;
- plough early after harvest and then flood for 2–3 weeks to rot dropped seed;
- rest the field from rice at least 1 year in 4;
- clean thoroughly all equipment, particularly harvesting machinery;
- in the longer term, introduce markers into the cultivars to be grown so that their foliage can be distinguished from that of wild rice, so allowing earlier roguing (e.g. anthocyanin pigments in foliage).

Apart from wild rice, other weeds for which standards exist in seed-certification schemes include: *Echinochloa* spp; *Panicum* spp; *Sorghum halepense*; *Cyperus rotundus*.

Crop layout

Upland rice can be handled in the same way as other rain-fed cereal crops. Swamp or floating rice can be sown *in situ* or can be transplanted. For a seed crop it is an advantage when sowing *in situ* to drill the seed in spaced rows, so making it easier for subsequent crop inspection or roguing. Transplanting gives a much higher multiplication rate, and is particularly useful in the early stages of a multiplication cycle: for direct sowing about 70 kg/ha are required, whereas for transplanting about 25 kg of seed will provide enough plants for 1 ha. Transplanting should be done in lines to make later operations easier.

Crop management
Seed crops do not differ in management requirements from crops for food production. Cultivars of subspecies *indica* will not tolerate excess nitrogen, which produces excessive vegetative growth; subspecies *japonica* can utilise higher application rates. It is important to ensure edequate supplies of phosphorus. Water management is very important, and it is generally recommended that maximum water depth should be maintained during flowering and into the early stages of seed development.

Harvesting
As the seed begins to ripen, water is drained from the field so that the ground is as dry as possible when the seed is ready for harvest. The seed is usually ready for harvest 4–6 weeks after flowering. The seed should be hard; moisture content is usually 18 per cent or above, but should not exceed 25 per cent when the crop is cut. Small areas can be cut by hand, the panicles being gathered in bunches and stacked in such a way that they will dry. If enough straw is left with the panicles the sheaves can be stooked in the field or stacked on racks; panicles with no straw have to be removed to a drying floor. Combines can be used for larger areas and are very suitable for upland rice; when used on irrigated areas, combines will usually be fitted with 'flotation' tyres to prevent the machine sinking in the soft soil. Seed at high moisture contents – above 25 per cent – may suffer injury which can impair germination; at moisture contents below 15 per cent there is increasing

Fig. 5.1 Rice fan (Photograph: Vogelenzang Andelst BV)

danger that an unacceptably high proportion of the rice seed will be dehulled. Small quantities can be threshed by hand by beating the panicles on a prepared clean floor, or a small thresher may be used (Fig. 5.1).

Seed after harvest

Drying

Seed should not be stored for any length of time if the moisture content exceeds 12 per cent in the more humid climates, although 14 per cent is possible in drier areas. For longer-term storage it is necessary to dry to below 8 per cent.

Seed cleaning

Essential seed-cleaning equipment is an air/screen cleaner, coupled with indented cylinders. A gravity separator can also be useful.

Seed treatment

The main seed-borne diseases are blast (*Piricularia oryzae*), brown spot (*Cochliobolus miyabeanus*) and Bakianae disease (*Gibberella fujikuroi*). Organomercurial seed treatments have proved most effective; alternatives for blast are Benomyl, thiabendazole, aureofungin or kasugamycin.

Hybrid cultivars

Hybrids based on cytoplasmic male sterility have been developed in China and grown extensively since 1975; they now occupy about 22 per cent of the Chinese rice crop area. A possible danger is that only one CMS source is currently in wide use thus rendering the hybrids vulnerable to disease epidemic although no difficulties have yet arisen in practice. It is only in the last few years that hybrid cultivars have been tested outside China. A difficulty for seed production is that pollen dispersal is naturally inadequate and has to be assisted to achieve a reasonable degree of cross-fertilisation. Cultural practices are designed to obtain a high panicle population by adjusting plant spacing and fertiliser and water application. Fertilisation is best at air temperature between 24 and 28 °C. Additionally pollen dispersal is assisted by removing the flag leaves to expose the panicles and by shaking the plants by beating with sticks or dragging a rope across the field; this requires much hand labour as the shaking must be done two or three times each morning during the period of pollen shedding. (Rice flowers usually open between about 09:00 and 11:30 hrs). It is important to achieve synchronisation of flowering between the male and female parents; the flowering period of the male parent may be extended by using three fraternal lines. To ensure that the desired hybrid is produced, seed crops are isolated by 50 to 100 m. Seed yields of up to 3 t/ha have been achieved. For hybrids to succeed in areas where hand labour is not available many of these procedures would have to be mechanised and work is in progress in the USA to develop suitable techniques. The possible yield advantage of hybrids over conventional cultivars is 20 to 40 per cent.

Oryza glaberrima Stend. Ploidy : diploid

This species, known as African rice, is the original cultivated rice in West Africa. According to Purseglove (1985) some cultivars exist, including some which are floating. It is sensitive to photoperiod. It has now mostly been displaced by *O. sativa*, but can stand rougher treatment, so that some is still grown in West Africa in districts where water is not controlled.

Hordeum vulgare L. *sensu lato*. Ploidy : diploid
Common name: barley

Classification of cultivars

Ear type	Two-row; Six-row
Seasonal type	Winter; Alternative; Spring
Intended use	Animal feed; Malt

Main distinguishing characteristics of cultivars

Plant growth habit	Erect; Prostrate
Hairiness of lower leaf sheaths	Absent; Present
Anthocyanin coloration of flag leaf auricle	Absent; Present
Intensity of this coloration when present	Weak; Strong
Time of 50 per cent ear emergence	Early; Late
Anthocyanin coloration in awn tips	Absent; Present
Intensity of this coloration when present	Weak; Strong
Glaucosity of ear	Weak; Strong
Plant height	Short; Long
Ear density	Lax; Dense
Awn length relative to ear	Shorter; Longer
Awn spiculation on margins	Absent; Present
Sterile spikelets	Parallel; Divergent
Rachilla hair type	Short; Long
Grain husk	Absent; Present
Anthocyanin coloration of lemma nerves	Weak; Strong
Spiculation of inner lateral nerves of lemma	Weak; Strong
Hairiness of ventral furrow of grain	Absent; Present

The seed crop

Isolation

Barley is normally self-fertilised, although some six-row cultivars of winter barley may cross-pollinate to an appreciable extent. The usual isolation distance is 2 m when there is no physical barrier at the edge of the crop. If cross-pollination is likely to be a problem, isolation should be 250–300 m. When it is desired to provide isolation for protection against infection by seed-borne diseases a distance of at least 50 m is required.

Previous cropping

Barley seed can remain viable in the soil for long periods. To guard against the risk of volunteer plants a 2-year interval between barley crops is desirable except when the same cultivar is to be grown and the seed used each year can be authenticated. Six-row cultivars are particularly difficult to eradicate from a field once they have been introduced.

Difficult weeds

In the EEC, standards are in force for the following:

Avena fatua, *A. sterilis* or *A. ludoviciana* – wild oats
Lolium temulentum – darnel
Raphanus raphanistrum – wild radish
Agrostemma githago – corn cockle

Particular care should be taken when using hormone weed-killers as they can distort the barley ears making off-type recognition difficult.

Crop layout

Seed crops can be sown with normal cereal drills; tramlines are an advantage to assist in crop inspection and roguing. In the early stages of multiplication, particularly in the pre-basic and basic seed stages, lower seed rates and wider spaces between the rows can increase the multiplication rate, although yield per unit area may be less. It is useful practice to cultivate the field borders to prevent ingress of grass weeds from hedgerows.

Crop management

Seed crops should be managed in the same way as crops for food production. Fertiliser application should be adjusted to minimise the risk of lodging. Pest and disease control should be practised to ensure that the seed is well filled. The use of straw shorteners is not recommended in seed crops since they may make it more difficult to assess cultivar purity.

Seed crop inspection

The best time to assess cultivar purity is after ear emergence when the seed has started to fill (growth stage 75–85).

Harvesting

Seed crops are usually combined. The seed should be hard and difficult to mark with the thumb-nail; the straw should have lost its green colour. Seed should not be harvested when moisture content is above 20 per cent as mechanical injury can occur; very dry seed (below 8 per cent) is also vulnerable.

Seed after harvest

Drying

Seed can be stored at 14 per cent moisture for sowing in the next growing season. For longer storage, moisture content should be reduced below 10 per cent; when it is desired to keep samples for long periods they should be dried below 8 per cent. The temperature of the drying air should not exceed 49 °C and when drying seed at higher moisture contents this should be reduced to 44 °C.

Seed cleaning
Essential equipment is the air-screen cleaner which may be coupled with indented cylinders; a gravity separator may also be useful. A de-awner, if not fitted to the combine or thresher, should be included in the seed-cleaning line.

Seed treatment
The main seed-borne diseases are leaf stripe (*Pyrenophora graminae*), net blotch (*P. teres*), covered smut (*Ustilago hordei*) and loose smut (*U. nuda*). The first three are controlled by organo-mercurial seed treatment or the more recently introduced treatments based on dithiocarbamates. Loose smut was controlled by hot-water treatment which was cumbersome to apply and often damaged germination. Recently, control has been achieved by formulations based on chemicals such as fenfuram. Organo-mercurials or their equivalent also give protection against many soil-borne fungi. Treatment with formulations based on benomyl or triforine and dimethylformahide can give seedlings protection against mildew (*Erysiphe graminis*) and treatment is also available against some soil-borne pests, for example against wire-worm with gamma-HCH and dimethylformahide.

Hybrid cultivars

There has been only limited success in the introduction of hybrid cultivars. Some male sterile lines exist, but there have been many problems in developing these to the stage where reliable hybrid cultivars can be produced.

Avena sativa L. Ploidy : hexaploid. Common name: oats

Classification of cultivars

Seasonal type	Winter; Alternative; Spring
Colour of grain (lemma)	White; Yellow; Brown; Grey; Black

Main distinguishing characteristics of cultivars

Plant growth habit	Erect; Prostrate
Hairiness of leaf blade	Weak; Strong
Time of panicle emergence	Early; Late
Flag leaf attitude	Rectilinear; Recurved
Hairiness of uppermost node on stem	Absent; Present
Orientation of panicle branches	Unilateral; Equilateral
Glaucosity of glumes	Weak; Strong
Glaucosity of lemma of primary grain	Absent; Present
Intensity of glaucosity of lemma of primary grain	Weak; Strong
Plant height (including panicle)	Short; Long
Hairiness of back of lemma of primary grain	Absent; Present

The seed crop

Isolation

The crop is self-fertilised so that a clear demarcation (physical barrier or gap of 2 m) is all that is required.

Previous cropping:

An interval of 2 years free from oats is desirable, except when the same cultivar is grown and the seed used to sow the crop can be authenticated.

Difficult weeds

The most difficult weed is the wild oat, which may be of any of the three species *A. fatua*, *A. sterilis* or *A. ludoviciana* depending upon the district where the seed crop is being grown. Whereas wild oat seed can be removed during seed cleaning from wheat or barley, it is virtually impossible to remove them from oat seed. It is therefore essential to ensure that oat seed crops are grown on ground free from wild oats. Wild oat seed can remain dormant in the soil for long periods and therefore if fields have become contaminated it is necessary to undertake a systematic and long-term eradication plan. Herbicides are available (e.g. Barban, Di-allate, Metoxuron), but hand roguing may also be necessary to eliminate wild oat plants from the crops preceding the oat seed crop; at least 4 years are normally required during which crops other than oats are grown to allow control of wild oat to be effective. There is little that can be done effectively to control wild oats in an oat crop. Apart from wild oats, other difficult weeds in oat seed crops are *Raphanus raphanistrum*, *Agrostemma githago* and *Lolium temulentum*.

Crop layout

There are no requirements for a seed crop which differ from those of a crop for grain production.

Crop management

Oats lodge very easily and seed crops should not be given too much nitrogen. There are no particular differences in management of a seed crop as compared with a grain crop apart from the details relating to quality control.

Seed crop inspection

To assess cultivar purity seed crops should be inspected after panicle emergence when the grain is just starting to fill. A later inspection when colour changes can be seen in the seed (lemma) may also be desirable.

Harvesting

Modern oat cultivars can be harvested satisfactorily by combine, although the seeds in an oat panicle tend to ripen less uniformly than do seeds of other cereals. The seed should be hard and difficult to mark with the thumb-nail. The straw tends to retain more green colour when the seed is ripe than does the straw of other cereals.

Seed after harvest

Drying

Seed moisture content should be below 16 per cent for short-term storage, and should be 14 per cent or less when seed is to be stored for some months.

For long-term storage moisture content should be below 10 per cent and samples should be reduced below 8 per cent before being placed in moisture-proof containers. The drying air temperature should not exceed 44 °C when the initial moisture content is high, but this can be increased to 49 °C as drying proceeds.

Seed cleaning

An air screen cleaner and indented cylinder are the main requirements for seed cleaning. The practice of clipping or polishing oat seed is not recommended: although it improves appearance and can also make seed flow more easily, it can also damage germination.

Seed treatment

See precautions mentioned under 'wheat' (p. 67). The main seed-borne diseases of oats and current treatments are as follows:

Brown foot rot and ear blight *Fusarium* spp.	Organo-mercury, guazatine, benomyl, thiophanate-methyl, triadomenol/fuberidazole/imazalil
Covered smut *Ustilago hordei*	Organo-mercury, carboxin/thiram, guazatine + imazalil
Leaf (stripe) spot *Pyrenophora avenae*	Organo-mercury, guazatine + imazalil
Loose smut *Ustilago avenae*	Organo-mercury, guazatine + imazalil, carboxin/thiram

Avena nuda L. Ploidy : hexaploid

The naked oat is not much grown. Yields tend to be well below those of conventional oats, although because the kernel is free of the lemma and palea the feeding value of the grain is high.

Avena byzantina K. Koch. Ploidy : hexaploid

The cultivated red oat is grown in the Mediterranean area, Australia and South Africa. Some other areas use cultivars which are hybrids between *A. sativa* and *A. byzantina*. They are handled for seed production in a manner similar to that described for *A. sativa*.

Triticosecale spp. Ploidy : hexaploid, but some octaploid cultivars. Common name: triticale

Classification of cultivars

Seasonal type Winter; Spring

Main distinguishing characteristics of cultivars

Plant growth habit	Erect; Prostrate
Time of ear emergence	Early; Late
Flag leaf attitude	Rectilinear; Recurved
Auricle pigment	Weak; Strong
Auricle hairs	Absent; Present
Anther pigment	Absent; Present
Glaucosity of plant at flowering	Weak; Strong
Hairiness of neck	Weak; Strong
Plant height (stem and ear)	Short; Long
Ear length	Short; Long
Ear shape	Tapering; Parallel
Ear density	Lax; Dense
Distribution of awns	Tip only; Upper half; Whole length
Length of awns relative to ear length:	Shorter; Longer
Grain coloration with phenol	Light; Dark

The seed crop

Isolation

Although triticale is largely self-fertilised, some cross-fertilisation can occur, particularly with certain cultivars. Depending upon the cultivar, therefore, isolation can be provided by a physical barrier or a gap of 2 m, or if out-pollination is a possibility an isolation distance of 250 m should be provided from other triticale crops; this should be increased to 300 m for the earlier generations.

Previous cropping

An interval of 2 years free from cereals – particularly triticale, wheat, rye and barley – should be allowed before growing a triticale seed crop; an exception can be made when the preceding crop is triticale of the same cultivar, grown from authentic seed.

Difficult weeds

There is as yet little information on weeds which may be peculiar to triticale crops, but it may be assumed that those which cause problems in wheat or rye will also cause difficulty in triticale.

Crop layout, crop management, seed crop inspection

Similar to that described for wheat.

Harvesting

The seed should be hard and difficult to mark with the thumb-nail; the straw will have lost its green colour. The seed does not shed easily, but in some cultivars the ears may break off at the neck if allowed to remain too long in the field before harvest. Shrivelled or poorly developed seed may be a problem in some cultivars.

Seed after harvest

Drying and seed-cleaning
Requirements are similar to those described for wheat.

Seed treatment
Triticale has so far been relatively free from seed-borne diseases although some loose smut (*Ustilago nuda*) has been reported; treatment similar to that described for wheat is possible. A major problem with triticale is ergot (*Claviceps purpurea*). Many ergots can be removed during seed cleaning, but it is very difficult to remove all and it is thus very important to maintain seed stocks free from ergot. There is no effective chemical control. For small seed quantities it is possible to float the ergots out of a seed sample by stirring it in a 20 per cent solution of common salt in water; the salt must be washed from the triticale seed and the seed dried immediately after treatment. Another alternative is to keep seed in store for a year before use, as ergots are normally short-lived; this is an added reason for avoiding taking a seed crop as a second triticale crop in succession. Field hygiene is essential and grasses growing at the edges of fields should be kept cut at time of anthesis to avoid ergot development which might give rise to additional sources of infection. Some cultivars of triticale may be resistant to ergot infection, but there is little precise information.

CROSS-FERTILISED CEREALS

Secale cereale L. Ploidy : mainly diploid, but some tetraploid cultivars available. Common name: rye

Classification of cultivars

Seasonal type	Winter; Alternative; Spring
Ploidy	Diploid; Tetraploid

Main distinguishing characteristics of cultivars

Seed; colour of aleurone layer	Light; Dark
Anthocyanin coloration of coleoptile	Absent; Present
Plant growth habit	Erect; Prostrate
Glaucosity of flag leaf sheath	Weak; Strong
Time of ear emergence	Early; Late
Glaucosity of ear	Weak; Strong
Hairiness of stem below ear	Weak; Strong
Plant height (stem and ear)	Short; Long
Ear length	Short; Long

The seed crop

Isolation

Rye is cross-fertilised. The OECD Cereal Seed Scheme (OECD 1977) requires an isolation distance of 250 m for crops producing certified seed and 300 m for earlier generations. These distances reflect the isolation requirements normally found in quality-control systems. Isolation between diploid and tetraploid cultivars is particularly necessary as infertility can result from pollination by a parent of a different ploidy.

Previous cropping

An interval of 2 years free from cereal crops is desirable before growing a rye seed crop. Rye is prone to shedding and therefore volunteer plants can be widespread in a subsequent crop. Because rye is cross-fertilised volunteer plants can cause additional problems by releasing pollen into the seed crop so giving rise to off-types in subsequent crops grown from the seed. Rye may follow rye when the crops are the same cultivar and the seed used to sow them can be authenticated.

Difficult weeds

Rye is a temperate cereal, and the weeds normally associated with wheat, barley or oats are also troublesome in rye. Wild oats (*Avena fatua*, *A. sterilis* and *A. ludoviciana*) are the most wide spread. *Raphanus raphanistrum*, *Agrostemma githago* and *Lolium temulentum* are other weeds listed by the EEC in the seed-certification scheme.

Crop layout and crop management

There are no requirements for a seed crop which differ from those for a food or feed crop. Spaces between rows wide enough to permit roguing can be left at regular intervals, and these are also a help at seed-crop inspection. Nitrogen should be used sparingly on seed crops since most cultivars are somewhat weak in the straw and lodge easily.

Harvesting

Rye is harvested in a similar manner to wheat or barley. The seed is ready for harvest when it is hard and difficult to mark with the thumb-nail; the straw will have lost most of its green colour. Threshing should not take place when seed moisture content is above 20 per cent as otherwise germination may be damaged. The straw is generally rather tough and may tend to cause blockages if not sufficiently dry. Very dry grain may also suffer mechanical damage and loss of germination.

Seed after harvest

Drying

The seed should be below 16 per cent moisture for temporary storage, and should be dried to below 14 per cent as soon as possible for keeping to the next sowing season. For longer storage, 8–10 per cent is desirable, and samples for long-term storage should be below 8 per cent. Drying should be slow, and the temperature of the drying air should not exceed 44 °C; when initial seed moisture content is below 16 per cent this can be increased to 49 °C.

Seed cleaning

An efficient air/screen cleaner is essential for cleaning rye seed. Indented cylinders and a gravity separator are also useful in the seed-cleaning line. Rye is very susceptible to damage and requires careful handling to maintain satisfactory germination.

Seed treatment

The smuts are the main seed-borne diseases. *Urocystis occulta, Tilletia caries* and *Ustilago tritici* can be controlled by organo-mercurial treatments. The most serious disease is ergot (*Claviceps purpurea*) which is widespread. The ergots are difficult to remove from seed during cleaning and seed growers should concentrate on growing a clean crop. Using year-old seed on clean fields is the best control, as ergot is short-lived and will die in store; it is also necessary to keep grasses in the field verges from flowering to prevent ergot from living on these alternate hosts. Small quantities of infected seed may be cleaned by floating the ergot off in a 20 per cent solution of common salt in water; the rye must be washed free of salt and dried immediately after immersion.

Hybrid and synthetic cultivars

Self-fertility exists in rye and it is possible to produce inbred lines for the production of F1 hybrid cultivars. Some hybrid cultivars have been produced but have not reached great commercial importance. Synthetic cultivars produced by interpollination of several inbred lines for a limited number of generations have also been tried. So far, seed-production experience with these types of cultivar has been limited.

Zea mays L. Ploidy : Diploid. Common name: maize, corn

The species is subdivided into the following types, which are sometimes given the status of botanical varieties:

Dent	The seed has a dent in the apex caused by shrinkage of soft starch.
Flint	The seed is filled, with no dent.
Flour	The endosperm is soft starch.
Sweet	The seed contains a large proportion of sugar.
Popcorn	The seed has a thick layer of hard starch surrounding soft starch at the centre of the endosperm.
Waxy	The starch in the seed is composed entirely of amylopectin, giving it a waxy appearance.

In agriculture, the types used are mainly 'Flint' and 'Dent', with a limited amount of 'Waxy' for special uses. 'Flour' is now rarely grown except in some local communities. 'Sweet' and 'Popcorn' are grown as specialist food crops, and sweet corn is regarded as a vegetable (George 1985).

Classification of cultivars

Breeding system	Open-pollinated; Synthetic; Inbred line; Hybrid
For an open-pollinated cultivar	Conventional; Composite
For a hybrid cultivar	Single-cross; Double-cross; Top cross; Other
Type of seed:	Flint; Dent; Waxy
Maturity	Early; Late
Photoperiod sensitivity	Insensitive; Intermediate; Sensitive

Main distinguishing characteristics of cultivars

Leaves	Not erect; Erect
Time of silk emergence	Early; Medium; Late
Anthocyanin coloration of silks	Weak; Strong
Closed anthocyanin ring at base of glume on tassel	Absent; Present
Plant height (including tassel)	Short; Long
Peduncle length at maturity	Short; Long
Ear coverage at maturity	Tip visible; Tip not visible
Colour of tip of grain	White; Yellow; Orange; Red; Black
Anthocyanin coloration of glumes of cob	Absent; Present
Ear length at maturity	Short; Long
Ear shape at maturity	Cylindrical; Conical
Number of rows of grain	Few; Many

The seed crop

Isolation

Maize is cross-fertilised and adequate isolation is therefore needed to prevent pollination from an undesirable source. (Fig. 5.2). The minimum isolation distance normally required is 200 m. A greater distance may be required for earlier generations in a multiplication cycle up to a maximum of 600 m. Isolation distances can be reduced when the outside rows of plants are not taken for seed; the number of rows to be discarded depends upon the size of the field and the isolation distance available (Table 5.1). For the production of hybrid cultivars the discard rows must be planted with the male parent. In some circumstances it is possible to arrange isolation by using difference in time of pollen shedding; natural differences between cultivars (including inbred lines) in the time from sowing to plants which shed pollen may be enhanced by adjusting sowing times. However, care is needed during hybrid seed production to ensure that the female silks become receptive at the time that the desired male parent is shedding pollen. As pollen shedding may extend over some weeks it is essential to ensure that the time interval is adequate.

Fig. 5.2 A maize seed crop

Table 5.1 Border rows to be discarded to achieve adequate isolation

Isolation distance available (m)	Number of rows to be discarded according to size of field (ha)						
	Less than 1.6	1.6–2.4	2.4–3.2	3.2–4.0	4.0–4.8	4.8–6.4	Over 6.4
175	3	3	2	2	2	1	1
150	5	5	4	4	3	3	2
125	7	7	7	6	5	5	4
100	9	9	8	8	7	7	6
75	11	11	10	10	9	9	8
50	13	13	12	12	11	11	10

Source: Technical Guideline for Maize Seed Technology, FAO, Rome, 1982

Previous cropping

Plants arising from shed seed are not normally a problem with maize, and crops may follow one another on the same field. However, there may be volunteer plants, especially when two crops follow one another within the same year; when volunteer plants are likely it is advisable to allow an interval of one crop of another species otherwise hand-pulling of the volunteer plants will be necessary.

Difficult weeds

Weed seeds are not normally a problem in maize seed. Usually weeds are not harvested with the maize, and if this should happen weed seeds can be removed relatively easily during cleaning. There are therefore no special measures which need to be taken for a seed crop; however, maize is very sensitive to competition, particularly in the early stages of growth, and cleanliness is therefore essential to obtain a satisfactory seed yield.

Crop layout

Maize seed is sown in rows with the seeds spaced in the rows. The distance between rows can be 50–100 cm with 75 cm the most usual; wider spacing may be used in the tropics. The optimum plant population depends upon the growing conditions and can vary from 25 000 to 100 000 plants/ha with 50 000 the most usual. To achieve 50 000 plant population on 75 cm inter-row spacing the distance between plants in the row should be 26 cm. For the production of hybrid cultivars it is necessary to plant the correct proportions of male and female parents. When inbreds are used as in single-cross hybrid production the usual ratio is four rows of female to two of male; other combinations are also possible, for example, two female to one male or three to one. It is essential to choose a male parent which sheds pollen when the female silks are receptive or to adjust sowing times of the two to ensure that this happens; the male parent should also be tall enough to ensure that pollen is dispersed over the crop. For double-cross hybrid production the male parent will normally be more vigorous and it is possible to have the female : male ratio at 6 : 2 or 8 : 2. When using cytoplasmic male sterility in the female parent it is necessary to use a restorer otherwise the progeny will also be sterile. This may be done in either of two ways: by mixing seed of the male sterile female parent with seed of that parent in which male sterility has not been introduced in the proportion 2 : 1; or by using a male parent which contains one or more specific restorer lines so that at least one-third of the plants grown from the resulting hybrid produce pollen which appears normal in all respects.

Crop management

Open-pollinated cultivars do not require management for seed production different from that applied in the production of a feed crop. For hybrid cultivar production the female parent must be emasculated before the silks become receptive to ensure that the desired cross takes place. When cytoplasmic male sterility is used it is only necessary to check at regular intervals that none of the female parent plants have produced tassels which are shedding viable pollen. For mechanical detasselling of the female parent it is necessary to ensure that action is taken at the correct time. The tassels on the female parent plants must be removed before they begin to shed pollen. This can be done by hand or by special machine. Most of the tassels are removed at the first pass, but it is necessary to check subsequently that none have been missed; usually two to three checks should be made at intervals of a day or two. Mechanical detasselling thus takes about a week to 10 days and it is essential that it be performed conscientiously and regardless of weather conditions, otherwise the desired hybrid will not be produced.

Seed-crop inspection

For maize seed production seed-crop inspection is particularly concerned to ensure that isolation is satisfactory, and for hybrid cultivar production that the female parent is not producing pollen. In both cases it is also necessary to check that cultivar identity and purity are satisfactory. When the seed crop follows a maize crop in the preceding season an additional check should be made to ensure that there were no volunteer plants present. The OECD Maize Seed Scheme (OECD 1977) provides that crops producing seed of hybrid cultivars shall be rejected if more than 0.1 per cent of off-type plants in the pollen parent have shed pollen, and if at the last inspection there are more than 0.1 per cent of off-type plants in the female parent; for open-pollinated cultivars the maximum permitted off-type plants is 0.5 per cent in basic seed and 1 per cent in certified seed. For detasselling, the OECD standards are applied when 5 per cent or more of the parent plants have receptive silks: crops are rejected which contain more than 1 per cent of female parent plants with tassels which have shed or are shedding pollen at any one inspection, or a total of 2 per cent or more for three inspections on different dates. Tassels, sucker tassels or portions of tassels are counted as shedding pollen when 50 mm or more of the central stem, the side branches or a combination of the two have their anthers extended from their glumes and are shedding pollen.

Harvesting

Seed should not be harvested when moisture content exceeds 25 per cent since mechanical damage may occur. For harvesting by hand, moisture content between 15 and 20 per cent is preferred. Before harvest the seed is hard and the plants have turned yellow. Machine harvesting is best performed with a corn-picker which picks the ears and dehusks them; combines which also separate the seed from the cob are not generally advisable because of the risk of damage to the seed, but if they are used less damage is likely to occur at moisture contents between 25 and 35 per cent (Purseglove 1985). When harvesting hybrid cultivars it is important that the male parent be removed from the field first; in most cases it can be harvested for forage immediately after pollen shedding is complete, but if allowed to set seed the male rows should be picked first and removed to a separate place while the plants should be shredded. Any inadvertent mixture of seed from the male parent will spoil the quality of the hybrid seed.

Seed after harvest

Drying

Maize ears for seed are usually dried in cribs constructed with wire mesh or similar sides to allow air to flow freely through. Cribs should be smaller in the more humid regions (Table 5.2). If cribs are not available the ears may be dried in the open or in an airy building, but must be protected from rain and intense direct sunlight (see Fig. 5.3). If heated air is used to dry ears its temperature should not exceed 42 °C. After drying, moisture content of the seed should not exceed 14 per cent, and when initial moisture content is high drying must be done slowly. However, it is important to maintain a good circulation of air through the ears. When ears are stored on floors or in drying bins the depth must not be too great.

Table 5.2 Crib dimensions for drying maize seed

Dimensions of crib (m.)			Volume (m^3)	Capacity (kg ears)	For use where air humidity is:
Length	Width	Height			
35.00	1.35	3.00	142	70 000	High
35.00	1.50	4.00	210	110 000	Moderate
35.00	2.00	3.00	210	110 000	Low

Source: *Technical Guideline for Maize Seed Technology*, 1982, FAO, Rome.

Shelling

After maize ears are dry (14 per cent moisture) they are easily shelled either by hand or by machine. Special shellers are available for seed designed to minimise damage. After shelling it is desirable to check the seed moisture content and to ensure that it is no more than 14 per cent for short-term storage; for longer-term storage a moisture content of 10–12 per cent is satisfactory, while for long-term storage of seed stocks 8–10 per cent is required. In the early stages of multiplication it is advisable to hand-pick the ears before shelling, discarding those which are not true to cultivar or are diseased or damaged.

Seed cleaning

After shelling it is desirable to pre-clean the seed to remove debris such as broken cobs or husks and dirt. Subsequent cleaning can usually be achieved by an efficient air/screen cleaner. The screens are used to separate the small, medium and large seed. The majority will be medium-sized which has passed

Fig. 5.3 Drying maize in Nepal

through a 12 mm (approx.) screen but is retained by a 6 mm (approx.) screen. When the seed is to be planted with a plate planter the plates differ according to the size grade of seed being used. Accordingly, seed is divided into 'flat' (passed by a slotted screen) and 'round' (retained by a slotted screen). Flats usually occur in the middle of the ear where the seeds are packed tightly together, with rounds at either end; larger seeds occur towards the base of the ear. Flats and rounds are further graded for size into 'large', 'medium' and 'small' using round-hole screens. These separations can be improved on a gravity separator, and sometimes an indented cylinder can be used to make separations according to seed length.

Seed treatment

Several fungi are seed borne and can cause loss when infected seed is sown. The main genera are *Pythium*, *Fusarium*, *Drechslera*, *Diplodia* and *Helminthosporium*. Seed treatment with organo-mercurial dressing is effective and alternatives are thiram, quinone, captan or benzimidazole. Captan also controls bacterial wilt caused by *Erwinia stewartii*.

Hybrid cultivars

In developed countries most of the maize area is planted with hybrid cultivars and many are now in use in the developing countries. Composites are also used in many developing countries. Inbred lines are usually produced by hand pollination in the first instance, followed by multiplication in well-isolated seed plots. Production of commercial hybrids may be achieved by the use of cytoplasmic male sterility or by mechanical detasselling; the former method may transmit disease susceptibility to the hybrid and male sterile lines must be carefully screened. The production of hybrids requires careful advance planning to ensure that the correct amounts of inbred seed are produced for the construction of the hybrid. Seed yields from inbred plants are usually very much reduced, generally no more than half normal. The area required to grow seed of the male parent will be only one-third of that required to grow the female, assuming a final planting ratio of one male to three female. The preparation of a planned multiplication is along the following lines:

1. Quantity of hybrid seed required divided by expected yield per ha from female parent = area (ha) of female parent required (A).
2. A multiplied by seed rate/ha = quantity of seed required to sow A (B).
3. B divided by expected yield per ha = area of female required in previous year (C).
4. C multiplied by seed rate/ha = quantity of seed required to sow C.

This process is continued until the quantity of seed is small enough to be produced from a specified number of plants. The quantities of male parent seed required at each stage are derived from the requirements for the female seed according to planting ratio.

Sorghum bicolor (L.) Moench Ploidy : diploid.
Common name: sorghum

This section is devoted to grain sorghums; those used for grazing or fodder are discussed in Chapter 6. House (1985) distinguishes five more or less distinct cultivated complexes each originating in a particular area. Harlan and de Wet (1972) distinguish five main races described by spikelet characteristics as follows:

Bicolor: Grain elongate, sometimes slightly obovate, nearly symmetrical dorso-ventrally; glumes clasping the grain, which may be completely covered or exposed as much as ¼ of its length at the tip; spikelets persistent.

Guinea: Grain flattened dorso-ventrally, sub-lenticular in outline, twisting at maturity 90 degrees between gaping involute glumes that are nearly as long to longer than the grain.

Caudatum: Grain markedly asymmetrical, the side next to the lower glume flat or even somewhat concave, the opposite side rounded and bulging; the persistent style often at the tip of a beak pointing towards the lower glume; glumes ½ of the length of the grain or less.

Kafir: Grain approximately symmetrical, more or less spherical, not twisting, glumes clasping and variable in length.

Durra: Grain rounded obovate, wedge-shaped at base and broadest slightly above the middle; the glumes very wide, the tip of a different texture from the base and often with a transverse crease across the middle.

There is considerable variability in many of the older cultivars of sorghum and several other classifications have been made by other authors (e.g. Purseglove 1985). Doggett (1965) referring to the earlier classification by Snowden in 1936 states 'the species boundaries are vague and there are all sorts of intermediate forms'. The above classification, as the author implies, is by no means absolute and we can expect many intermediate types since the classes are all inter-fertile. In addition, there are many F1 hybrid cultivars, particularly in the USA and India, the majority of which are based on female lines from milo (durra) and male from kafir.

Classification of cultivars

Breeding system	Open-pollinated; Synthetic; Inbred line; Hybrid
Maturity	Early; Medium; Late
Plant height	Short; Long
Photoperiod sensitivity	Insensitive; Intermediate; Sensitive

Main distinguishing features of cultivars

The following morphological characteristics are adapted from House (1985):

Plant colour	Pigmented; Tan
Stalk juiciness	Dry; Juicy
Stalk sweetness	Insipid; Sweet
Leaf midrib colour	Colourless (white); Dull green; Yellow; Brown
Ear compactness	Lax; Compact
Ear shape	Elliptic; Oval; Broomcorn
Glume colour	White; Yellow; Brown; Red; Purple; Black; Grey
Seed covering by glumes	Uncovered; Covered
Awning at maturity	Awnless; Awned
Seed colour	White; Yellow; Red; Brown; Buff
Endosperm colour	White; Yellow

Seed-coat lustre	Not lustrous; Lustrous
Presence of subcoat	Absent; Present
Seed form:	Dimpled; Not dimpled
Twinning of seeds	Single; Twin

The seed crop

Isolation

Sorghum is largely self-pollinated, but also out-pollinated to the extent of 5–10 per cent. For purposes of practical seed growing, therefore, it has to be regarded as cross-fertilising, and suitable isolation must be provided for seed crops. For hybrid cultivars controlled cross-pollination is required and isolation for them is obligatory. The minimum distance required is 200 m with up to 400 m isolation when markedly contrasting types are being grown – for example between a grain sorghum cultivar and Johnson grass or other fodder sorghum or between a grain sorghum and broom-corn. There can be considerable variation within the seed crop in time of flowering and hence of pollen release; consequently isolation in time is difficult to achieve and is not generally recommended although possible in some instances.

Previous cropping

One year free from sorghum should be allowed before planting a seed crop.

Difficult weeds

Seed crops should be maintained weed free to avoid loss of yield. *Striga* spp. are a particular problem in this respect in Africa. Some of the wild species of sorghum and the cultivated forage sorghums can create particular problems in seed crops; they can release pollen which may fertilise some of the plants in the seed crop, and their seeds are difficult to remove from harvested seed of a grain sorghum crop. Control measures in the seed crop are difficult and costly and therefore clean fields should be chosen for seed crops.

Crop layout

Sorghum is usually planted in rows at a relatively low seed rate to obtain plants spaced in the row. Distances between rows can vary from 46 to 90 cm and between plants in the row from 15 to 60 cm. For machine sowing 75 cm between rows and 10 cm between plants in the row is satisfactory. For the production of hybrid cultivars it is necessary to plant the correct proportion of female : male parents. This is normally 6 : 2. The outer rows should all be male (pollinator) line and there should be four of these. It is also important that the two lines flower at the same time so that pollen is being shed when the female plants are receptive. It is quite usual for the period from sowing to flowering to be shorter for the male line than for the female, and it is then necessary to sow the female line some 10–14 days before the male; the seed grower must obtain guidance from the plant breeder on this point and may find it necessary to conduct small-scale trials before undertaking large-scale production. It is also possible to influence time of flowering by applying nitrogen or irrigation water. Application of nitrogen (e.g. as a 4 per cent spray of urea) or of irrigation water can bring forward the flowering date of one line by up to 7 days. To do this the crop layout should be planned in advance so that irrigation water or nitrogen can be applied to one or the other line as required. Date of flowering can be forecast by examining the floral initials about 4 weeks after crop establishment by carefully slitting the stem of a sample of plants from each line; at this stage if the initials in one line are smaller than those in the other, action should be taken.

Crop management

Apart from the points discussed above, a seed crop requires the same basic management as a food or fodder grain crop. Additional nitrogen should be applied shortly before flowering, preferably as a side dressing at least 5 cm away from the plants. However, this will only be effective if water is not limiting.

Seed-crop inspection

It is generally necessary to inspect sorghum seed crops at least twice, and hybrid cultivars three times. The critical times are during flowering, to check isolation and in hybrid varieties that no female plants are shedding pollen; and at maturity, when seed colour can be observed to confirm cultivar identity.

Harvesting

According to House (1985) seed is physiologically mature when moisture content falls to 30 per cent, although Chopra (1982) gives 25 per cent. At this stage, however, the seed is still soft and liable to damage if handled roughly. In most cases, therefore, it is preferable to wait for the seed to harden and for moisture content to fall to below 15 per cent before harvesting. At this stage plants will have lost most of their lower leaves and, depending on the cultivar, the upper leaves may or may not have turned yellow. Sorghum is prone to sprouting in wet weather, so it is preferable to start harvest at the first opportunity in difficult climates. Many crops are harvested by hand, and for older cultivars this has the advantage that the later heads can be left longer on the plants to mature by cutting on two or three occasions. When hand harvesting it is possible to cut the whole plant or to remove only the seed heads. In either case it will generally be necessary to leave the harvested material to dry before threshing, either on the field or on a prepared drying floor. Larger areas can be combined, and this method is particularly suited to the more uniform and shorter-strawed cultivars. When harvesting hybrid cultivars it is essential to remove the male rows first and to ensure that all heads from these rows are removed from the field before starting to harvest the hybrid seed. If the male rows are combined it is essential to follow the combine and to remove any missed heads.

Seed after harvest

Drying

After hand harvesting the seed must be dried to 12 per cent moisture or below before threshing. Seed heads can be sun-dried on a prepared floor, but should not be spread more than 20 cm deep and must be carefully turned at frequent intervals. Threshed seed can be dried using forced air at a temperature not exceeding 40 °C.

Seed cleaning

Sorghum seed is very vulnerable to storage pests and moulds and must not be placed in store if moisture content is above 10–11 per cent. All stores must be kept scrupulously clean and the same applies to drying and threshing floors. It pays to use an insecticide on floors before using them for sorghum. Pre-cleaning of threshed seed, either by winnowing at the threshing floor or in the cleaning plant, is necessary to remove broken straw and other debris. Final cleaning can be achieved on an efficient air/screen cleaner; sometimes a gravity separator may also be used.

Seed treatment

The main seed-borne diseases are the smuts (*Sphacelotheca* spp.) for which treatment with organo-mercurials is most effective. Alternative but less effective treatments are sulphur, choranil, thiram and carboxin. Some fungi which attack seedlings (for example *Drechslera* and *Fusarium*) are also stopped by seed treatment with organo-mercurials, while thiram and chloranil are effective against *Cercospora sorghi* and *Gleocerospora sorghi*.

Hybrid cultivars

The production of hybrid cultivars depends upon the use of cytoplasmic male sterility, and hybrids are usually F1. In the production of the female (male sterile) line it is necessary to use a 'maintainer line' which will permit seed to be taken from the male sterile line which will in turn produce male sterile plants. When it is desired to take the seed of the hybrid cultivar a 'restorer line' must be used which ensures that the F1 seed harvested from the male sterile line will produce plants which reproduce normally. This process is fully explained by Chopra (1982). Calculations of seed requirements are similar to those described for hybrid maize.

Pennisetum glaucum (L.) B. Br. emend. Stuntz. Ploidy : diploid

Pennisetum glaucum is the name used in ISTA (1984). However, both IBPGR (1981) and Purseglove (1985) use *P. americanum* (L.) Lecke, which is given in ISTA (1984) as a synonym, as is *P. typhoides* (Burm.) Staph et Hubb. The most used common names are pearl millet and bulrush millet.

Classification of cultivars

Breeding system	Open pollinated; Synthetic; Inbred line; Hybrid
Maturity	Early; Medium; Late
Photoperiod sensitivity	Insensitive; Intermediate; Sensitive

Main distinguishing characteristics of cultivars

Culm thickness	Thin; Thick
Shape of inflorescence	Cylindrical; Conical; Spindle; Club; Dumb-bell; Lanceolate; Oblanciolate; Globose

Bristle length	Shorter than seed; Longer than seed
Exposure of seed	Seed exposed in spikelet at maturity; Not exposed
Seed shape	Obovate; Lanceolate; Elliptical; Hexagonal; Globular
Seed colour	Ivory; Cream; Yellow; Grey; Brown; Purple; Black

The seed crop

Isolation

Pearl millet has two types of flowers: the hermaphrodite flowers mature first and are followed by a second burst of anthesis from male flowers which have only stamens. The plant is thus mainly cross-pollinated and adequate isolation is required for the seed crops. The minimum isolation distance normally required is 200 m, but for the earlier generations up to 1000 m is often considered necessary. More than 200 m would also be required between strongly contrasting cultivars – for example, those used for grain as opposed to those used for grazing or fodder. Because of the extended period of pollen release, isolation in time is not recommended.

Previous cropping

An interval of 1 year free from millet should be allowed before a seed crop is taken to reduce the risk of volunteer plants occurring.

Difficult weeds

Clean crops will give the best seed yield. *Striga* spp. are not so great a problem in millet as they are in sorghum, but the crop is susceptible.

Crop layout

Pearl millet is usually planted in rows for a seed crop as this enables subsequent operations to be performed more easily. The distance between the rows should be somewhat higher than that for sorghum since millet tillers more profusely: 60–100 cm between rows and 10–20 cm between plants in the row. For the production of hybrid cultivars the proportion of female to male parents must be satisfactory – usually 6 : 2, but other proportions may be necessary. The outer rows in a field should all be male (pollinator) line and usually this surround will be four rows deep.

Crop management

A seed crop does not require management different from that applied to a food or grain crop. The crop responds well to nitrogen and a seed-bed application will hasten early growth. Bird damage is a particular problem with pearl millet and seed crops require particular attention as they reach maturity.

Seed crop inspection

Seed crops should be inspected twice; once to check isolation after emergence of the inflorescence but before anthesis; and once later when cultivar characteristics are more easily observed. Hybrid cultivars should be inspected additionally to check that no female plants are shedding pollen.

Harvesting
Hand harvesting can begin when the seed reaches physiological maturity, usually when moisture content is around 25 per cent, but it is better to wait until moisture content falls to 15 per cent. Panicles then have to be dried to below 12 per cent moisture before threshing. With some cultivars which produce many side tillers, hand harvesting on more than one occasion may be worth while. Pearl millet is prone to sprouting in the ear, so harvest should not be delayed, especially in areas where unsettled weather may be expected. Larger areas may be harvested by combine. Whether harvesting by hand or by combine special care is needed with hybrid cultivars to ensure that seed is taken only from the female parent, and to achieve this the male rows should be removed from the field first.

Seed after harvest

Drying
Seed heads must be dried to 12 per cent moisture before threshing. This can be achieved by sun-drying either in the field or on a prepared drying floor. In the latter case the seed heads should not be spread more than 20 cm deep. When using forced-air drying the temperature of the drying air should not exceed 40 °C.

Seed cleaning
Pearl millet is less vulnerable in store than sorghum, but care is still needed to avoid storage pests; scrupulous cleanliness is required at all times. Pre-cleaning of threshed seed is desirable, either by winnowing or on a pre-cleaner at the cleaning plant. Subsequently, cleaning can be completed on an efficient air/screen cleaner; occasionally a gravity separator will also be used.

Seed treatment
There is no evidence that seed treatment is necessary for pearl millet. Most of the diseases to which it is prone are either soil- or airborne and can be combated best by other measures.

Hybrid cultivars

The production of hybrid cultivars of pearl millet depends upon cytoplasmic male sterility as described for sorghum.

Eleusine coracana (L.) Gaertu. Ploidy : tetraploid. Common name: finger millet

Classification of cultivars

Finger millet is an important food crop in parts of Africa and India, but has generally received less attention than pearl millet. Consequently the cultivars are less well defined and described. Purseglove (1985) refers to a classification by K. L. Mehra, which divides the cultivars into two groups:

1. African highland types with long spikelets, long glumes, long lemmas and with grains enclosed within the florets.
2. Afro-Asiatic types with short spikelets, short glumes, short lemmas and with mature grains exposed out of the florets.

Main distinguishing characteristics of cultivars

Purseglove (1985) lists the following:

Plant height	Dwarf; Tall
Colour of vegetative organs	Green; Purple
Tillering	Little; Much
Type of inflorescence	Spikes straight and open; Spikes incurved and closed; Spikes branched resembling a cockscomb
Length of spikes	Short; Long
Number of spikelets per spike	Few; Many
Length of spikelet	Short; Long
Tightness of packing of grain	Lax; Dense
Length of glumes	Short; Long
Seed colour	White; Orange red; Deep brown; Purple/black

The seed crop

Finger millet is grown in a similar way to pearl millet. There are no hybrid cultivars, and although the crop is largely self-fertilised (about 1 per cent cross-pollination is common) some isolation should be provided. In the areas where finger millet is grown, wild species are common and may cause problems either by releasing pollen or as mechanical mixtures in the seed crop. Finger millet stores well, being less prone to storage pests than either pearl millet or sorghum.

6 GRASSES

Unlike the cereal crops, grasses are not usually grown for their seed, but to produce leafy forage for livestock. In consequence, seed crops usually require management which is different from pasture or hay and most seed crops are now grown with a definite seed market in mind. The brief descriptions of management practices in this chapter cover the main points, but this is a developing field of study, especially in so far as weed control by the use of herbicides is concerned. For some species, specialised harvesting equipment is being devised, but is still in the experimental stage; this applies particularly to the grasses used in the tropics. Another development is the possible use of chemicals for three main purposes: to shorten the flowering stem so as to make harvesting easier; to use desiccants on the mature crop to reduce moisture content before harvest; and to prevent seed shedding of ripe seed by the use of resins or polymers. None of these practices has yet reached the stage where they can be recommended for general use, but experimental work is in progress which could provide additional tools of great benefit to the grass-seed grower.

Most grass species are cross-fertilised and in consequence the plant-to-plant variation within a cultivar is relatively high. Therefore it is possible that certain environmental influences or management practices may favour seed production from one kind of plant while being unfavourable for others. This has particular relevance when seed production takes place in an area far removed from the place where the parent plants of the cultivar were selected and the original breeders' seed produced. Much work has been done to elucidate this problem, particularly in the USA where herbage seed is produced in the West, where climate is suitable for seed production, of cultivars for use in the eastern states and in other countries where climate is more suitable for pasture or hay. In general the results show that growing seed for one generation in a different environment does not have a significant effect; cultivars differ in their stability and some knowledge of this is needed in interpreting this generalisation (see e.g. Kelly and Boyd 1966).

Irrigation is practised in the production of some grass seed and can be beneficial. It allows production to take place in areas where the climate is very suitable for seed maturation, but where seed yields might otherwise be low because of earlier drought stress. Irrigation can be used during the establishment phase to ensure a good strong plant and to encourage the development of plenty of early tillers for seed production; water stress later encourages the production of flowering stems in a more uniform way, while subsequent further irrigation encourages the production of plump seed. Irrigation timing has to be integrated with nitrogen application to ensure that excessive leaching does not occur; nitrogen dressings may have to be applied more frequently in smaller doses. There are some species which do not respond to irrigation but there is little experimental evidence; generally the species which are adapted to drier conditions (e.g. tall fescue among the temperate grasses) respond less to irrigation than those which are less drought resistant.

TEMPERATE GRASSES USED MAINLY IN AGRICULTURE

These grasses are adapted to a temperate climate. There are annual and biennial forms, but most are perennial. The biennials and perennials respond to lengthening days and temperature to change from the vegetative to the reproductive phase. They also usually form tillers in the autumn which make the greatest contribution to seed yield in the following year.

Lolium spp. – ryegrass

Although the ryegrasses are normally divided into Italian ryegrass – short-lived cultivars which are annual, biennial or persist no longer than 3 years – and perennial ryegrass – cultivars able to persist for several years – the dividing line between these two is not clear cut, and in practice there is a continuum from the very short-lived annual Westerwold ryegrasses to the very persistent perennials. There are numerous hybrids between *L. multiflorum* and *L. perenne* and it can be difficult accurately to assign a particular cultivar to a particular species. The position is further complicated by the fact that both diploid and tetraploid cultivars are in use. For the seed grower these complications pose problems, particularly in relation to previous cropping of the seed-production field and in providing satisfactory isolation.

Lolium multiflorum can normally be distinguished from *L. perenne* by the presence of awns on the outer paleas. Additionally, the rachilla is usually flattened in *L. multiflorum* but oval in *L. perenne*; the shoots of *L. multiflorum* are rounded, but those of *L. perenne* flattened; the auricles are more prominent in *L. multiflorum*. Within *L. multiflorum*, cultivars which are short-lived and which produce a seed crop in the year of sowing when sown in early spring in northern Europe are called 'Westerwolds'. The fluorescence test has also been used to distinguish short-lived cultivars: seeds are germinated in the dark on non-fluorescent paper and the seedlings are then examined under ultraviolet (UV) light; cultivars with a high proportion of seedlings which fluoresce are generally those which are less persistent. However, this association is not absolute and some cultivars of *L. perenne* also contain a high proportion of fluorescent seedlings.

Hybrids between *L. multiflorum* and *L. perenne* may exhibit more or less of the characteristics of either species depending upon the selection pressure imposed by the breeder.

Cultivars which are tetraploid have seeds which are almost double the size of those which are diploid; moisture content of the plant as a whole is 1–2 per cent higher, so that drying is slower. Some tetraploids are less persistent than their diploid equivalents, but the more recent tetraploid cultivars have

Lolium multiflorum Lam. Common name: Italian ryegrass, including Westerwold

Classification of cultivars

Seeding in the sowing year	Annual (Westerwold); Biennial (Italian).
Ploidy	Diploid; Tetraploid.

Main distinguishing characteristics of cultivars

Time of inflorescence emergence	Early; Late
Plant growth habit (at inflorescence emergence)	Erect; Prostrate
Flag leaf length (at inflorescence emergence)	Short; Long
Flag leaf width (at inflorescence emergence)	Narrow; Wide
Stem length, including fully expanded inflorescence	Short; Long

The seed crop

Only one seed harvest is taken from Italian ryegrass seed crops. A second crop usually gives a greatly reduced yield and it is impossible to prevent excessive volunteer plants from shed seed, leading to difficulties in the maintenance of cultivar purity.

Isolation
Italian ryegrass is cross-fertilised and pollen dispersion is by wind. For isolation purposes *L. multiflorum* and *L. perenne* have to be treated as if they are one species. The OECD Herbage and Oil Seed Scheme (OECD 1977) requires an isolation distance of 200 m for small fields (2 ha or less) and 100 m for larger fields when the seed produced is intended for further multiplication; these distances are reduced to 100 m and 50 m when the seed produced is to sow crops for fodder or other use. Diploid and tetraploid cultivars will not cross-fertilise, but the pollen may cause infertility, so reducing seed yields; consequently an isolation distance of 50 m is recommended.

Previous cropping
Most grasses can persist in a field and it is advisable to allow an interval of at least 2 years free of grasses with similar seed size before sowing an Italian ryegrass seed crop. In some seed-certification schemes intervals of 3 or 4 years are specified. The use of a defoliant such as paraquat to kill established grass plants may be a technique which could shorten the interval needed between seed crops. In general, spring crops such as cereals or root crops provide better opportunities to clean fields than autumn-sown crops; the latter tend to encourage the build-up of volunteer grasses and grass weeds.

Difficult weeds
Grass weeds are the most difficult to control, and their seeds the most difficult to remove after harvest. These include: *Avena fatua* and *A. ludoviciana* (wild oat); *Alepecurus myosuroides* (black grass); *Agropyron repens* (couch); *Poa annua* (annual meadow-grass) *P. trivialis* (rough-stalked meadow-grass); *Holcus lanatus* (Yorkshire fog); *Bromus* spp. (bromes); *Lolium temulentum*

(darnel). Grass weeds should be controlled in the seed bed. Glyphosphate can be applied in a preceding crop of cereals, and controls many broad-leaved weeds as well as the grasses. On stubbles or as a pre-emergence treatment, paraquat is generally effective. After establishment, ethofumesate will control many grass weeds and also volunteer cereals, cleavers and chickweed; it should be applied in late autumn/early winter (middle of October to middle of December in the UK) when the soil is moist; the ryegrass seed crop should be growing well, and have at least two to three leaves. Broad-leaved weeds should be controlled by the use of suitable chemicals such as MCPA, 2,4-D, mecoprop, ioxynil and bromoxynil. These herbicides are usually safe to use when the Italian ryegrass has reached the three- to four-leaf stage in the autumn, and may also be applied in the spring up to 4 weeks before the start of inflorescence emergence; if applied at the wrong time they can cause sterility and consequent loss of seed yield.

Crop layout

Seed crops should preferably be sown in rows which are close together (inter-row space 10 cm, maximum 20 cm). Broadcasting may be used in wetter areas, but is not generally recommended as the seed is not placed at an even depth. Westerwold ryegrass is sown in early spring without a cover crop, and taken for seed in the same year. Crops of Italian ryegrass may be under-sown in a cereal crop in the spring, or sown in autumn without a cover crop. Autumn sowing can take place at the end of August, and should be completed by mid-September in northern Europe. For diploid cultivars, the seed rate is 12 kg/ha in good conditions and 16 kg/ha when the seed bed is less satisfactory; for tetraploids the corresponding rates are 16 and 22 kg/ha. The higher rates are used when broadcasting.

Crop management

So far as the major nutrients – calcium, phosphorus and potassium – are concerned, the requirements are approximately the same as for wheat, with perhaps a slightly greater emphasis on potassium. The soil pH must be above 5 and preferably around 6. Applications of 45–50 kg/ha of phosphorus and 55–60 kg/ha of potassium will be required on most reasonably fertile soils. Nitrogen is applied as necessary to stimulate growth at critical times. When crops are under-sown or when direct sown in autumn under difficult conditions an autumn dressing of 20–50 kg/ha will encourage growth, particularly if the crop is to be grazed. The objective is to stimulate tillering so as to establish a good base for the seed crop. In spring, an additional 100–175 kg/ha will be needed, depending upon the field conditions: under wetter conditions less nitrogen is advisable to reduce the risk of excessive lodging, and total spring application should not exceed 125 kg/ha. Initial application should be in mid-March and some authors suggest that a single application at this time is effective; others, however, favour a split application with about one-third in mid-March and the remainder later after defoliation. Autumn grazing in the year of establishment is advantageous, and it pays to remove excessive growth from ungrazed crops before midwinter.

It is possible to take an Italian ryegrass seed crop without spring defoliation, but there are two disadvantages: lodging may occur before anthesis, which can lead to a poor seed set and there may be excessive vegetative growth to deal with at harvest. Defoliation mitigates against these two possibilities and has the added advantage that the value of the grazing or silage offsets some of the costs of growing the crop. However, defoliation must be timed correctly or decreased seed yield will result. Grazing or an early silage cut are possible, but in either case the field should be closed up for seed by mid-April and certainly before the end of April in the UK. Later defoliation causes either fewer or smaller inflorescences, both of which lead to reduced seed yield. Defoliation should not take place when the field is excessively wet and there is danger of poaching.

Seed-crop inspection

The best time for inspection is just at the beginning of inflorescence emergence, when the cultivar characteristics are most easily observed and when there is still the possibility of correcting any deficiencies in isolation. Standards for cultivar purity are usually expressed as the maximum numbers of off-types permitted per unit area because it is very difficult to distinguish individual plants within the crop and so impossible to assess on a percentage basis.

Harvesting

The choice lies between swathing and picking up later and direct combining. In climates where good weather at harvest is normal, swathing is preferred, but where drying of the harvested seed is usually necessary, direct combining is better. As with most grass species seed maturation is soon followed by shedding; the seed grower also has to contend with a crop where the inflorescences will be at different stages of ripening. Generally, however, the greatest weight of seed is yielded by the longer inflorescences born on the older tillers and it is best to harvest when the majority of these are ripe rather than waiting for the later shorter inflorescences to mature. Seed is usually doughy when ripe and the green colour is beginning to fade; in some cultivars anthocyanin coloration may intensify at this stage. The stem also begins to yellow. Another good guide to ripeness is to check the seed moisture content. At this stage moisture loss can be rapid so that frequent checks are necessary. The crop is sampled by taking inflorescences at random from several places in the crop and placing them in a polythene bag; the whole sample is then threshed, ensuring that all seed is removed, and the moisture content determined. For swathing to begin the moisture content should be 45 per cent or just under for tetraploid, and 43 per cent for diploid cultivars. Moisture contents should fall a further 5 per cent before direct combining (i.e. 40 per cent for tetraploids and 38 per cent for diploids). In good weather the moisture content usually falls by 2–3 per cent per day at this time; the effect of rain causing a temporary apparent increase in moisture content must be discounted.

Seed after harvest

Drying

In areas with good weather it is best to dry the crop in swath and to pick up with a combine when the seed moisture content is low enough for safe storage – 14 per cent or below. Where field drying is less certain and when the crop is direct combined, the harvested seed must be dried immediately:

germination can be damaged very quickly if damp seed is put into store, and tetraploid cultivars are particularly vulnerable. When using on-floor drying, ventilation should begin immediately and moisture content must be reduced below 20 per cent within 5 days. Advice from the National Institute of Agricultural Botany, Cambridge, is that seed depth should not exceed 55 cm when seed moisture content is over 35 per cent (floor area required is then 8 m^2/tonne). The required airflow per tonne for seed at moisture contents 35, 40 and 45 per cent is 16.2, 21.54 and 27.8 m^3/min. No heat is used in the early stages, and only very little in the final stages when the air may be reduced to 65 per cent humidity. It is important to ensure that the top surface also dries to below 20 per cent within 5 days. When using continuous-flow, flat-bed driers, the seed should not be more than 15 cm deep to avoid overdrying the bottom layer and to avoid large moisture gradients in the seed. Maximum temperatures for the drying air are given as 38, 49 and 54 °C for seed at moisture contents of 45, 40 and 35 per cent. It is important to cool seed which has been dried in warm air before placing it in store.

Seed cleaning

An efficient air/screen cleaner will normally produce a good-quality sample; in some instances a gravity separator may be needed to remove certain impurities.

Diseases and pests

The most serious seed-borne disease is blind seed disease (*Gloeotina temulenta*) and this is more prevalent in perennial than in Italian ryegrass. Ergot (*Claviceps purpurea*) is also seed borne, but is not generally a problem. There are no seed treatments available to combat blind seed disease, although small quantities of seed can be immersed in hot water (50 °C) for 30 minutes, followed by cooling and drying. The best remedy is to use clean, disease-free seed; there are some resistant cultivars and seed which has been stored for more than 2 years is usually free of the disease, which will not survive in dry seed. Field sanitation is also important. In Oregon, USA, straw burning *in situ* has been very effective in controlling disease but is now questioned on environmental grounds (Youngberg in Hebblethwaite 1980). There are several leaf and stem diseases which attack seed crops, but the effects are not generally considered serious enough to warrant the use of chemical fungicides; resistant cultivars should be grown in areas where such diseases are prevalent. Damage to young seedlings by frit fly can be combated by spraying chlorpyrifos; for slugs, the use of pellets of metaldehyde or methiocarb at sowing time is advised. After inflorescence emergence aphids can cause damage and if they occur should be controlled by spraying an aphicide. Midges can also cause damage is some instances.

Lolium perenne L. Common name: perennial ryegrass

Classification of cultivars

Ploidy	Diploid; Tetraploid
Intended use	Grazing or forage; Amenity (sports fields, etc.)

Maturity group (inflorescence emergence)	Very early; Early; Intermediate; Late; Very late

(These groups represent arbitrary divisions on a continuous scale, and cultivars occuring on the borderline between two groups may be classified differently in different countries.)

Main distinguishing characteristics of cultivars

Inflorescence emergence	Very early; Early; Intermediate; Late; Very late
Flag leaf length	Short; Medium; Long
Flag leaf maximum width	Narrow; Medium; Wide
Length of longest stem (including inflorescence)	Short; Medium; Long
Length of inflorescence	Short; Medium; Long

The seed crop

Cultivars of early perennial ryegrass are usually harvested for seed in the year after establishment only. The later cultivars can be managed so as to produce a reasonable yield from a second harvest and occasionally a third harvest year is possible. However, the greatest seed yield is obtained from the first harvest year.

Isolation

Requirements were discussed under *L. multiflorum*. However, with *L. perenne* the range of flowering times between cultivars is greater and in some cases it may be possible to arrange isolation in time rather than space between a very early and a very late perennial where the difference in flowering time can be a month.

Previous cropping, difficult weeds and crop layout

These were discussed under *L. multiflorum* and the same considerations apply to *L. perenne*. However, seed rates slightly lower than those quoted will suffice for *L. perenne*.

Crop management

Perennial ryegrass is managed in a similar manner to Italian ryegrass during establishment and during the first harvest year. However, care must be exercised during spring grazing: seed crops of early cultivars should be closed up earlier – generally by mid-March or at the latest by the end of March in the UK. Late cultivars can be defoliated later, but crops should be closed up before the end of April. In some drier areas it may be preferable not to defoliate in spring provided growth is not excessive. If the crop is to be taken for a second or third seed harvest the field should be cleaned up as soon as possible after harvest either by burning or by removing all cut material and trimming any patches which were not cut closely enough. Nitrogen dressings are applied at the same times as in the first year, but the rates can be increased by about 30 per cent. On soils low in phosphorus or potassium additional applications may be needed, but it is generally best to correct such deficiencies before sowing the crop. Autumn and spring grazing or defoliation follows the same pattern as in the first year. On some soils it will pay to roll the field when it is closed up to press down any stones which might interfere with harvest.

Seed crop inspection, harvesting, drying and seed cleaning
These follow the same pattern as for Italian ryegrass. Ripeness in the early perennials resembles Italian ryegrass, but in the late perennials the stems remain greener when the seed is ready for harvest. Moisture content of the seed at harvest is also some 2 per cent lower in perennial compared with Italian ryegrass.

Diseases and pests
Blind seed disease occurs more frequently in *L. perenne* than in *L. multiflorum*; otherwise the major diseases and pests are as outlined for *L. multiflorum*.

Lolium × Boucheanum Kunth. Common name: hybrid ryegrass

Hybrids between *L. multiflorum* and *L. perenne* may resemble the one or the other, or may be intermediate in their characteristics. For seed growing they should be treated as either Italian or perennial, depending upon which they resemble most closely. Both diploid and tetraploid cultivars are available.

Lolium rigidum Gaud. Common name: Wimmera ryegrass

This annual species is grown for seed in the south-east of Australia. Production methods are similar to those for Westerwold ryegrass.

Festuca pratensis Huds. Common name: meadow fescue

Classification of cultivars

Time of inflorescence emergence	Early; Late

Main distinguishing characteristics of cultivars

Length of flag leaf	Short; Medium; Long
Width of flag leaf	Narrow; Medium; Wide
Length of longest stem (inflorescence included)	Short; Medium; Long

The seed crop

Seed-crop requirements are similar to those for *L. perenne*, except that as the seed is generally somewhat larger, seed rates should be similar to those quoted for *L. multiflorum* (diploid). Defoliation is beneficial in the autumn, but grazing or cutting in the spring must be completed early, before the end of March otherwise lower seed yields will result; it is best to omit spring defoliation unless growth is excessive. There is little evidence on moisture content at harvest; crops are ready for swathing when the seed is turning brown, but retains some green colour and the stems also are beginning to lose their green colour. Early cultivars are more suitable for direct combining that those maturing later, which are usually much more leafy. Blind seed disease does not occur in *F. pratensis*.

Hybrids between *Lolium* and *Festuca*

Hybrids between *Lolium* and *Festuca* have been produced, and are known to occur in nature. However, such hybrids have generally proved to be poor seed producers and none has yet been developed commercially.

Alopecurus pratensis L. Common name: meadow foxtail

There are very few cultivars of this species and very limited areas are grown for seed, mostly in central Europe and the north-west of the USA. The seed heads tend to ripen over an extended period and much seed is lost through shedding. Best results are obtained where the growing season is short as this tends to cause the crop to ripen more uniformly. Seed crops can be grown in a similar manner to those of perennial ryegrass and can be combined direct as the flowering stems are long enough to enable the cutting table to work above the lower dense vegetative growth.

Dactylis glomerata L. Common name: cocksfoot, orchard grass

Classification of cultivars

Time of inflorescence emergence	Early; Late

Main distinguishing characteristics of cultivars

Length of flag leaf	Short; Medium; Long
Width of flag leaf	Narrow; Medium; Wide
Length of longest stem, inflorescence included	Short; Medium; Long

The seed crop

Cocksfoot seed crops are usually harvested for seed for 3 years. Seed crops may thus occupy a field for 4 years, an establishment year plus 3 harvest years.

Cocksfoot is cross-fertilised by wind-borne pollen. Isolation requirements, previous cropping and difficult weeds are as discussed under Italian ryegrass.

Crop-layout

The best seed yields are usually obtained when seed crops are established in rows spaced widely apart; depending on growing conditions, inter-row spacing may be from 45 to 60 cm, the narrower spacing being used when moisture is not limiting. Spacing wider than 60 cm may be used under very dry conditions. Seed rate is approximately 8 kg/ha in wide drills.

Crop management

Spring sowing is preferred, with or without a cover crop. If a cover crop is used it must not be too heavy or it will suppress the seedling cocksfoot plants too much. First harvest seed yields will generally be higher from crops sown without a cover crop, and the choice will depend upon the relative prices of the cocksfoot seed and the cover-crop produce. Inter-row cultivations are beneficial in suppressing weeds and in providing a good environment in which the seedlings can grow. Cultivation should not be too deep, and it is essential to avoid moving soil on top of young seedlings. Cultivation is especially beneficial after the removal of a cover crop so as to loosen the soil between the rows. However, inter-row cultivation is relatively expensive, and for most weed control the use of herbicides should be considered (see 'Difficult weeds' in section on *L. multiflorum*). Unless autumn growth is excessive, defoliation before the first seed harvest offers no advantage. Excess leaf may cause some loss of seed yield if it is killed by winter frost; it should be removed late in the year but before the end of January when the crop should be closed up for seed. After the first seed harvest the crop should be cleaned of excess herbage and in dry areas may be burned to advantage. Inter-row cultivation is beneficial if the soil has become compacted; however, care should be taken not to damage the roots of the cocksfoot by cultivating too deep. Any subsequent defoliation should be completed in December/January, and is only needed if growth is excessive. Spring cultivations are not normally necessary. Lodging is not usually a problem with cocksfoot, and the crop responds well to nitrogen, given that the calcium, phosphorus and potassium status is satisfactory. Some nitrogen may be needed in the sowing year to stimulate quick establishment of the seedlings; subsequently an application of 100–150 kg/ha in February/March each harvest year is recommended in the UK. There is some evidence that very early flowering cultivars may respond better to a split dressing, half or more in autumn and the remainder in spring.

Seed-crop inspection

The best time for inspection is at the start of inflorescence emergence. For crops sown in wide rows the sample areas for detailed examination should be a predetermined length of row, including half the inter-row space on either side of it.

Harvesting

The crop may be swathed and subsequently picked up or it may be combined direct. The former method is preferred in areas where good weather will prevail throughout the period the crop is in swath, and the seed can be combined at approximately 14 per cent moisture; however, when this method of harvesting is to be followed the crop should not be grown in rows spaced too widely, otherwise it is difficult to keep the swath off the ground.

Swathing can take place when the seed is about 44 per cent moisture; at this stage most of the seed will be light brown, but some will still be greenish, and stems below the inflorescence will have turned yellow to brown. Some seed shedding may have occurred. For direct combining the seed moisture content should be 30 per cent or below, which will normally be about 10 days after the swathing stage; however, combining may have to be started earlier if seed shedding becomes excessive.

Seed after harvest

Drying and seed cleaning.
These are as described for Italian ryegrass.

Diseases and pests
Ergot (*Claviceps purpurea*) can occur in cocksfoot, but is not generally serious. Choke (*Epichloe typhina*) causes damage to the crown of the plant, and often occurs in older stands (third harvest year onwards); the disease may be seed borne and seed should not be harvested from badly infected fields. Powdery mildew (*Erysiphe graminis*) is often prevalent in cocksfoot seed crops; two or three tridemorph or benomyl sprays in autumn followed by one or two in spring have given good control in areas where the disease causes crop damage. There are several other leaf and stem diseases which occur on cocksfoot, but so far control measures have not been shown to be worth while. Cocksfoot moth (*Glyphipterix cramerella*) can cause damage in some seasons, particularly in older stands; burning stubbles after harvest and generally clearing field boundaries of roughage destroys the sites where larvae hibernate. Midges may also attack cocksfoot occasionally; eggs are laid in the developing inflorescences and the larvae feed on the seed when they can be controlled by insecticides.

Phleum pratense L. Common name: timothy

Classification of cultivars

Time of inflorescence emergence	Early; Intermediate; Late

Main distinguishing characteristics of cultivars

Length of flag leaf	Short; Medium; Long
Width of flag leaf	Narrow; Medium; Wide
Length of longest stem, inflorescence included	Short; Medium; Long

Cultivars of this species are hexaploid.

The seed crop

The management described for cocksfoot applies equally to timothy except for the following points:

Crop layout and management

Timothy succeeds well with narrower inter-row spacing than cocksfoot, and solid stands are achieved from spacing of 10–20 cm. However, weed control is often easier in wide-spaced rows of 45–60 cm. Seed rate is approximately 6 kg/ha in wider spacing and 8 kg/ha for close spacing. When crops are sown without a cover crop it is often advantageous to sow a marker plant with the seed (lettuce or mustard) sufficient to provide a plant every metre or so to show up the rows earlier. Spring defoliation can generally take place about 2 weeks later than for cocksfoot, since timothy is usually later to form inflorescences.

Harvesting

The crop is ready for swathing when the seed has changed colour to grey with a brownish tinge and the inflorescence is losing seed from the tip. Seed threshes more easily from the swath after a period of alternate wetting and drying – light rain or dew followed by sun and wind (Fig. 6.1). Some growers prefer to direct combine the crop first with a wide clearance at the concave so as to knock out only the ripest seed, and then to return to pick up and rethresh the crop at a normal combine setting.

Diseases and pests

Timothy is attacked by leaf and stem diseases but they rarely do sufficient damage to warrant treatment. Timothy fly (*Amaurosoma* spp.) may occur in

Fig. 6.1 Timothy in swath

some seasons, but is so sporadic that treatment has not been recommended. Treatment against aphids or midges may be required in some seasons.

Phleum bertolonii DC. Common name: timothy

This species is treated in all respects like *Phleum pratense*, of which it is the diploid equivalent. The plants are smaller and have a creeping habit. The seed is used also for amenity purposes and to provide a bottom grass in orchards.

Festuca arundinacea Schreb. Common name: tall fescue

Classification of cultivars

Time of inflorescence emergence	Early; Late

Main distinguishing characteristics of cultivars:

Length of flag leaf	Short; Medium; Long
Width of flag leaf	Narrow; Medium; Wide
Length of longest stem (inflorescence included)	Short; Medium; Long

The seed crop

Seed-crop requirements are similar to those outlined for cocksfoot. Tall fescue normally differentiates inflorescences early, and benefits from autumn nitrogen. Defoliation should be practised with care and must on no account be done too late in the spring. Harvest ripeness is similar to that described for meadow fescue; the seed sheds easily and harvest losses can be excessive if harvest is delayed. After harvest seed crops should be cleaned up and excess herbage removed; in some areas burning the stubble improves subsequent seed yields. Tall fescue is quite difficult to establish, but can produce many seed crops (some authors suggest as many as fifteen seed harvests from one stand).

Bromus inermis Leyss. Common name: soft brome; bromegrass; smooth brome

Two regional groups of cultivars of soft brome are recognised. The group from the southern USA has more vigorous underground root-stocks than the northern group and produces earlier spring growth. In Europe some cultivars have been developed, but detailed descriptors are not available.

Seed-crop requirements are similar to those outlined for tall fescue. The seed is chaffy and difficult to thresh and clean. In consequence it is difficult to handle and may present problems at sowing time. Seed also does not store satisfactorily for prolonged periods of more than a year.

Other *Bromus* spp.

These include the following:

Bromus arvensis L.	Field brome
Bromus carinatus Hook et Arn.	California brome
Bromus sitchenensis Bong.	
Bromus willdenowii Kunth.	Rescue grass; Prairie grass.

Cultivars of all these species are available in Europe and of the last also in the USA and New Zealand.

Arrhenatherum elatius (L.) Beauv. ex J. S. et K. B. Presl. Common name: tall oat grass

There are a few cultivars of this species, mainly in Europe. Seed yields are higher from crops grown in rows spaced wide apart. The seed sheds very easily and should be combined when the stem turns yellow and before the top half of the panicle has shed; this can be tested by walking into the crop and striking a number of panicles on the palm of the hand – if the seed in the top half can be threshed in this way the crop is ready for combining. The crop is difficult to thresh and the long, twisted awns on the dorsal pales make it difficult to handle. However, if the seed is threshed too hard and the pales removed the naked caryopsis tends to loose germination. The species is said to respond well to nitrogen.

TEMPERATE GRASSES USED MAINLY FOR AMENITY PURPOSES

The grasses in this section are used mainly for lawns, golf-courses, sports fields and other amenity or play areas. Some are also used in agriculture, sometimes in special situations, for example, creeping red fescue for hill grazings. There are also some grasses from the preceding section, particularly ryegrass, which are used for amenity areas such as football fields or on roadside verges; special cultivars may be selected for this purpose, but seed-growing technique does not differ in principle from that outlined previously.

When growing seed for use on amenity areas the seed grower must exercise particular care to ensure that pure seed is produced. Many cultivars are highly specialised and have been selected to provide a colour or other feature for use in particular situations. It is therefore necessary to ensure that there is no contamination either with other cultivars of the same species, or with seeds of other species, especially those which are coarser and do not form a good turf: for example, seed of red fescue must be absolutely free from seed of perennial ryegrass as even one or two ryegrass plants would spoil a red fescue turf. The fields selected for seed production must therefore be totally free from all volunteer grasses, whether these be weeds or cultivated forms.

Poa spp. meadow-grass

The meadow-grasses form an important group for amenity purposes and are used extensively to provide turf giving a hard-wearing, fine surface. They are also used in agriculture, particularly as pasture grasses.

Poa pratensis L. Common name: Smooth-stalked meadow-grass; Kentucky bluegrass

Classification of cultivars

Ligule pubescence	Not pubescent; Pubescent
Leaf sheath colour in young stage	Green; Red coloration
Upper leaf surface pubescence	Not pubescent; Pubescent

Main distinguishing characteristics of cultivars

Hairs on margin of leaf sheath	Absent; Present
Hair tuft on leaf sheath near top	Absent; Present
Leaf width	Narrow; Medium; Wide
Hair fringe at junction of leaf blade and sheath	Absent; Present
Hairs on lower leaf surface	Absent; Present
Culm length at maturity	Short; Medium; Long
Panicle colour	Mainly white; Mainly red
Rachis shape at junction with lower branches	Straight; Curved
Collar of rachis at junction with lower branches	Split; Closed
Angle of rachis branches	Inclined upwards; Horizontal; Drooping
Size of seed	Small; Medium; Large
Date of panicle emergence	Early; Medium; Late

The seed crop

Isolation
Poa pratensis is generally apomictic, but there are some cultivars which contain plants which are more or less sexual, when some degree of cross-fertilisation may occur. Seed growers should take advice from the plant breeder as to the isolation requirement for a particular cultivar. Most seed-certification schemes, however, assume apomixis and require only physical separation of the seed crop from possible sources of contamination with a gap of 3 m or wider.

Previous cropping
Poa spp. are generally widespread and it is important to select a field which has been cleaned of grass weeds, usually over a period of 2 years. Various herbicides can be used for this purpose provided the crop rotation is arranged so that crops are grown which permit their use. Seed crops of *Poa* spp. should be separated by an interval of at least 2 years – longer for seed crops of earlier generations.

Difficult weeds

Reference has already been made to the problems caused by grass weeds. *Poa pratensis* establishes slowly and the preparation of a weed-free seed bed, followed by a programme of weed control during crop establishment is essential. It is necessary to combat both broad-leaved weeds which hinder seedling growth and grass weeds which are competitive and cause further problems later if allowed to produce seed. Bromoxynil is reported by Canode (in Hebblethwaite 1980) to control annual broad-leaved weeds in established stands; perennial broad-leaved weeds can be treated with dicamba. On young plants which have reached the tillering stage 2,4-D or MCPA can also be used, and for earlier application, dinoseb. Grass weeds should be eliminated before the crop is sown, using a 'stale seed-bed' technique and propham or paraquat to kill grass weeds as they appear before sowing. After sowing, inter-row spraying with paraquat can be used, and for small numbers of grass weeds hand roguing may be necessary.

Crop layout

Poa pratensis responds well when sown in rows 30 cm apart. Spacing above 60 cm may reduce seed yield. Seed rate is generally low and can be about 1 kg/ha. Seed crops are normally sown in spring without a cover crop. If a cover crop is used it is advisable to use a crop such as winter oil-seed rape which can be harvested early to allow satisfactory development of the *P. pratensis* seedlings before autumn begins. A possible technique to aid establishment is to sow the *Poa* seed under a protective covering of charcoal about 2.5 cm wide. This allows the use of overall sprays of herbicides with the emerging *Poa* plants protected. Special equipment has been developed in Oregon for this (Youngberg, in Hebblethwaite 1980).

Crop management

Poa pratensis requires fertile conditions. Calcium, phosphorus and potassium status should be satisfactory. Seed-bed nitrogen is desirable to aid establishment, although under a cover crop it is advisable to delay application until after its removal. After the sowing year nitrogen is applied in the spring as early as is practicable. In Denmark 40–60 kg/ha has been recommended, but in the USA responses have been found from heavier applications up to 135 kg/ha, especially on older stands. There is no advantage from splitting the nitrogen application. *Poa pratensis* is very sensitive to shading from excess herbage in the seed crop. Cover-crop debris must be removed immediately, and where there is excess herbage from the *Poa* plants a light grazing will do no harm. However, the main seed yield will come from tillers formed early in autumn, and nothing should be done which might inhibit strong healthy tillering at this time. Three or four successive seed harvests can be obtained from a crop provided the field is not allowed to become sod-bound. When seed harvest is complete it is essential immediately to remove all excess herbage. Best results are generally obtained by burning the straw and stubble; an even, rapid burn leaves the field with exposed soil and reduces shed seed and weed seed. However, burning is disliked because it causes pollution from smoke, and mechanical removal of excess herbage may have to be used. Chopping the straw and stubble as close to the ground as possible and blowing the chopped material into trailers with a flail harvester has proved the most effective. An alternative is to remove the straw and then to burn the remaining stubble with a flame machine such as is used for removing tarmac from roads; in Oregon special 'field sanitisers' have been used experimentally for this purpose (Youngberg, in Hebblethwaite, 1980). However, fuel costs for this method are likely to be high. In some areas sod-bound seed-production fields have been rejuvenated by shallow ploughing, subsequently rolling the field and allowing the *P. pratensis* to grow up between the furrows. While this works well for production of less specialised seed, it is not generally recommended when specific cultivars are being produced because of the possibility that volunteer plants or weed grasses may grow up.

Seed-crop inspection
The best time for inspection is after emergence of the inflorescence, but preferably when the early emerging plants can still be distinguished.

Harvesting
The most usual method is to swath and pick up later with a combine. For swathing, the inflorescences will have changed colour to yellow or brown and the seed will be firm with a moisture content of 28 per cent. Picking up from the swath can take place when the seed has dried to 14 per cent moisture or below. The seed is difficult to thresh and may require double threshing; in Oregon combines have been adapted for this purpose.

Seed after harvest

Drying and seed cleaning
Poa pratensis is difficult to handle because the seed is enclosed in the pales which are hairy at the base; these hairs tend to entwine so that the seeds become clustered and it is necessary to break up these clusters so as to get the seed to flow freely. Seed harvested from the swath will normally be dry enough to pre-clean and store for some time. If, however, seed moisture is above 14 per cent it must be dried immediately or spread out thinly in a well-ventilated store where it can be turned frequently. Temperature of the drying air should not exceed 49 °C for moisture content of 30 per cent and may be increased by 5 °C as the seed dries. The seed is normally cleaned on an air/screen cleaner.

Diseases and pests

Diseases and pests do not usually attack *P. pratensis* seed crops sufficiently to warrant treatment. In some years insect attack may require treatment with an appropriate insecticide.

Other *Poa* spp.

These include (in order of importance):

P. trivialis L.	Rough-stalked meadow-grass; Rough blue grass
P. nemoralis L.	Wood meadow-grass; Wood blue grass

P. palustris L.	Swamp meadow-grass; Fowl blue grass
P. compressa L.	Canada blue grass
P. ampla L.	Big blue grass

Festuca rubra L. Common name: red fescue; Chewing's fescue; creeping red fescue

Chewing's fescue is distinguished from creeping red fescue by the absence of rhizomes. Creeping red fescue may have slender or strong rhizomes.

Main distinguishing characteristics of cultivars

Ploidy	Diploid; Tetraploid; Hexaploid; Octoploid
Anthocyanin in leaf sheath	Absent or very weak; Weak; Medium; Strong
Time of inflorescence emergence	Very early; Early; Medium; Late; Very late
Length of flag leaf	Very short; Short; Medium; Long; Very long
Width of flag leaf	Narrow; Medium; Wide
Stem length, inflorescence included	Very short; Short; Medium; Long; Very long
Length of inflorescence	Very short; Short; Medium; Long; Very long

The seed crop

Isolation
Red fescue is cross-fertilised and therefore requires adequate isolation of 200 or 100 m for fields of 2 ha or less and 100 m or 50 m for larger fields; in each case the greater distance is required when seed is being produced for further multiplication.

Previous cropping, difficult weeds, crop layout and management
The requirements of red fescue are similar to those described for *Poa pratensis*, although seed rates are higher because of the larger seed. Seed-crop management is also similar although, since red fescue is usually early to change from the vegetative to the reproductive phase, spring defoliation, if attempted, should be early. In the USA response to spring nitrogen has been rather less than for *Poa*, and up to 112 kg/ha is used (cf. *Poa* 135 kg/ha). In the UK spring nitrogen may cause excessive leaf growth with no seed yield advantage and greater benefits are found from autumn applications.

Seed-crop inspection
This should be somewhat earlier than for *Poa* so that isolation can be checked before anthesis.

Harvesting
This is similar to *Poa*, but the seed is relatively easy to thresh and handle. The seed does not shed before being ready for cutting and usually the moisture content is 25 per cent for swathing.

Seed after harvest

Drying and seed cleaning
These follow the same pattern as for *Poa*, but the seed flows freely and can be handled in the normal way, similar to ryegrass. Some insects attack red fescue but are generally kept in check by stubble burning.

Other *Festuca* spp.

Festuca ovina L. *sensu lato*	Sheep's fescue; fine-leaved sheep's fescue; hard fescue
Festuca heterophylla Lam.	Shade fescue

Seed production of these species is similar to that described for *F. rubra*. However, hard fescue in particular does not respond to stubble burning and the post-harvest residue should therefore be removed by chopping and carting away.

Agrostis tenuis Tibth. Common name: brown top; Colonial bent-grass

Main distinguishing characteristics of cultivars

Ploidy	Diploid; Tetraploid; Hexaploid
Plant growth habit after establishment	Erect; Medium; Prostrate
Leaf width	Narrow; Medium; Wide
Rhizomes	Absent; Present
Stolons	Absent; Present
Leaf colour	Pale; Medium; Dark
Time of inflorescence emergence	Early; Medium; Late
Plant growth habit at maturity	Erect; Medium; Prostrate
Flag leaf length	Short; Medium; Long
Flag leaf width	Narrow; Medium; Wide
Length of longest stem (inflorescence included)	Short; Medium; Long

The seed crop

Isolation
Brown top is cross-fertilised and requires isolation as described for *Festuca rubra*.

Other management practices
These are as described for *Poa pratensis*, although brown top probably

requires slightly less nitrogen. In Oregon, brown top has given some of the greatest responses to stubble burning.

Other *Agrostis* spp.

A. gigantea Roth	Red top
A. stolonifera L.	Creeping bent
A. canina L. subsp. *canina*	Velvet bent

These do not differ from *A. tenuis* in management requirements except that *A. stolonifera* and *A. canina* do not tolerate stubble burning: post-harvest residue should be chopped and removed.

PRAIRIE GRASSES

These grasses are normally associated with areas having a typically continental climate with relatively hot summers and cold winters.

Agropyron spp. Common name: wheat-grass

Ten species of *Agropyron* are listed in OECD (1984). These are:

A. desertorum (Fisch. ex LK.) Schult.	Standard crested wheat-grass
A. cristatum (L.) Gaertn.	Fairway crested wheat-grass
A. dasystachum (Hooker) Scribn	Northern wheat-grass
A. elongatum (Host) Beauv.	Tall wheat-grass
A. inerme (Scribn et J. G. Smith) Rydb.	Beardless wheat-grass
A. intermedium (Host) Beauv ex Banmg.	Intermediate wheat-grass
A. riparium Scribn et J. G. Smith	Streambank wheat-grass
A. smithii Rybd.	Western wheat-grass
A. trichophorum (LK.) K. Richter	Pubescent wheat-grass
A. trachycaulum (LK.) Malte ex A. F. Lewis	Slender wheat-grass

All of these are cross-fertilised except *A. trachycaulum* which is self-fertilised. Six cultivars are listed for *A. intermedium* but only one or two for each of the other species. No standard descriptors for cultivars of the wheat-grasses have been published.

The seed crop

Isolation
The cross-fertilised species require the usual isolation distances, i.e. 200 m for fields of 2 ha or less, or 100 m for larger fields when the seed to be produced is to be multiplied again, otherwise 100 m or 50 m. The self-fertilised *A. trachycaulum* should be isolated from other crops by a physical barrier or a strip 3 m wide.

Previous cropping

A 2-year interval between crops of wheat-grass is normally required, with longer intervals for the earlier generations. However, in some instances this interval can be shortened when a programme of chemical weed control has been undertaken.

Difficult weeds

For the seed grower the most difficult weeds in wheat-grass seed crops are grassy weeds such as *Agropyron repens* (couch-grass) or *Avena fatua* (wild oat) because the seeds are extremely difficult if not impossible to remove after harvest. Seed-production fields must therefore be thoroughly cleaned of these weeds before the seed crop is sown, and should be kept free during seed production, if necessary by hand-roguing.

Crop layout

The best seed yields are obtained from crops sown in rows spaced fairly wide apart (60 cm), with wider spacing being used in drier areas when irrigation is not available. Seed rates vary with the species and row spacing; rates of 2–3 kg/ha have been successful for *A. intermedium* sown in rows 60 cm apart.

Crop management

Weed control is most economically achieved by the use of suitable herbicides, with hand roguing of small numbers of grass weeds when necessary. Canode (in Hebblethwaite 1980) found no advantage from extensive inter-row cultivation and recommended only a light working with a rotary hoe to break the soil crust in the spring to provide a mulch. Nitrogen as a single application of 112 kg/ha in early spring gave better seed yields, especially in the fourth harvest year. The removal of old growth and straw immediately after harvest by grazing, mowing or burning is necessary, but other defoliation is not.

Seed-crop inspection

The usual time for inspection is at the time of inflorescence emergence when isolation can also be checked and if necessary corrected before anthesis.

Harvesting

Most wheat-grasses shed seed as they ripen and therefore direct combining must not be delayed. Those which are least troublesome in this respect are *A. dasystachum*, *A. intermedium* and *A. trichophorum*. For swathing, the seed is ready for harvest when it reaches the soft to hard dough stage. The combining stage is slightly later when the seed is hard; at this stage seed should knock out easily when the inflorescence is struck against the palm of the hand. When seed is grown in very wide rows it may be difficult to prevent the swath from lying in the inter-row space where it does not dry so well and is difficult to pick up with the combine. For combining direct it is possible to cut high so as to leave the bulk of the vegetation for removal later. Threshing is generally easy, but it is necessary to reduce the air-blast from the fan to the cleaning shoe to avoid blowing good seed over the back.

Seed after harvest

Drying

When picking up from the swath, seed moisture content will normally have been reduced below 14 per cent and the seed can therefore safely be stored. Direct combining may give seed with a higher moisture content and there will probably be green debris mixed with it; it should immediately be spread out to dry on a floor with good ventilation over it and turned frequently until the moisture content is reduced to a safe level.

Seed cleaning

An efficient air/screen cleaner will normally be all that is required to produce a good sample. Some species are awned and the seed may require de-awning. In most species the harvested seed will contain unbroken spikelets or seed clusters which should be broken up to enable the seed to flow easily in a seed drill.

Andropogon spp. Common name: bluestem

Three species of *Andropogon* are listed in OECD *List of Cultivars Eligible for Certification* (OECD 1984). These are:

A. geradii Vitm.	Big bluestem
A. hallii Hach.	Sand bluestem
A. scoparius Michx.	Little bluestem

These species are cross-fertilised. Three cultivars are listed for *A. geradii* and one each for the others. No descriptors for cultivars of *Andropogon* have been published.

The seed crop

Requirements for *Andropogon* are similar to those of *Agropyron*. Seed production is very dependent on the availability of adequate soil moisture during the culm-extension and seed-filling phases; also temperature at this time should not be excessively high. Bluestem is generally vigorous and will soon fill the inter-row spaces; some inter-row cultivation is desirable, but defoliation in the spring of the harvest year should be avoided. The crop responds well to nitrogen. The seed usually matures late in the season. At each joint on the raceme is a fertile, sessile spikelet and an infertile, stalked spikelet. A sample of the fertile spikelets should be checked to ensure that seed is developing; if it is not, it may be preferable to cut the crop early for hay. The harvested seed will consist of the fertile spikelets together with the stalks of the infertile spikelets; awns are usually present. Consequently the seed is difficult to handle and usually requires to be further broken up in a de-awner before cleaning. The debris, including the infertile spikelets, can be removed in an air/screen cleaner.

Elymus spp. Common name: wild rye

Only *E. junceus* Fisch. (Russian wild rye) is included in OECD (1984), with five cultivars. *E. canadensis* L. (Canadian wild rye) is also grown for seed in limited quantities. Both species are cross-fertilised. No descriptors for cultivars of *E. junceus* have been published.

The seed crop

Requirements for *Elymus* are similar to those of *Agropyron*. The seed crops are best established in wide-spaced rows and inter-row spacing up to 2 m has been used in very dry conditions. The crop responds well to nitrogen. Russian wild rye seed sheds very easily and it is therefore desirable to start harvest early when the seed is still in the dough stage for swathing, or somewhat later when it is beginning to harden for direct combining. Canadian wild rye sheds seed less easily, allowing some latitude in harvesting. The seed is awned and will require de-awning during cleaning to enable it to flow freely.

Boutelua spp. Common name: grama

Only *B. oligostachya* (Nutt.) Torr. ex A. Gray (side oats grama) is included in OECD (1984) with six cultivars. The species is cross-fertilised. No descriptors for cultivars have been published.

The seed crop

Requirements for *Boutelua* are similar to those of *Agropyron*. The seed matures on the spike irregularly and it is necessary carefully to time harvest to obtain the maximum yield of good seed. Direct combining is the best method, but the harvested seed will contain some immature seed and green vegetative debris; it must therefore be handled carefully and spread immediately on a well-ventilated floor where it can be turned frequently.

Phalaris spp. Common name: canary grass

There are two species included in OECD (1984), each with six cultivars.

Ph. arundinacea L.	Reed canary grass
Ph. aquatica L.	Harding grass; Phalaris

These species are cross-fertilised. No descriptors for cultivars have been published.

The seed crop

Requirements for *Phalaris* are similar to those of *Agropyron*. The seed tends to mature first at the tip of the spike and then progressively towards the base;

some loss of seed from the tip will occur before the optimum harvest date is reached which is when about half the seeds have turned brown. Cultivars have been selected to minimise seed shedding.

Other species

The following species are included in OECD (1984). For seed production, the requirements are similar to those of *Agropyron*.

Buchloe dactyloides (Nutt.) Engelm	Buffalo grass
Eragrostis curvula (Schrader) C. G. Nees	Weeping love-grass
Sorghastrum nutans L. Nash ex Small	Indian grass
Panicum virgatum L.	Switch grass
Stipa viridula Trin.	Green needle-grass
Trisetum florescens (L.) Beauv.	Golden oat-grass

TROPICAL GRASSES

Grasses grown in the tropics and to some extent subtropical areas are adapted to produce seed under short-day conditions and some are inhibited from seed production in longer days. Some species prefer humid conditions and others are adapted to drier areas. Some tropical grasses are propagated vegetatively, but this section includes only those normally grown from seed.

One of the difficulties encountered in the production of seed of tropical grasses is the irregularity with which viable seed is produced. Boonman (1971a) uses the criterion of yield of 'pure germinating seed' and points out that this may be less than the yield of clean seed, some of which may be infertile. Against this, seed is usually produced two or three times in a year, depending upon the availability of moisture; seed is also usually produced in the year of sowing, unlike the perennial temperate grasses. Yields at each harvest are low: Boonman (in Hebblethwaite 1980) mentions commercial yields in the order of 100 kg/ha with a quality of 25 per cent pure germinating seed. In those tropical countries where labour is plentiful, seed heads can be harvested by hand and this partly overcomes the difficulties caused by an extended ripening period. This is caused by prolonged heading both within and between plants and prolonged flowering within each head. To some extent flowering can be concentrated by achieving higher early tiller density (so suppressing late tillers) through closer row spacing and added nitrogen, but this is not always effective.

Chloris gayana Kunth. Common name: Rhodes grass

Seven cultivars are included in OECD (1984). Some are diploid and some tetraploid. Descriptors for cultivars have not been published.

The seed crop

Isolation

Rhodes grass is cross-fertilised and requires the usual distance specified in the OECD Herbage and Oil Seed Scheme (OECD 1977), i.e. 200 m, or 100 m for small fields (2 ha or less) depending upon whether or not the seed is for further multiplication, and 100 m or 50 m for larger fields. Diploid and tetraploid cultivars will not inter-cross but some infertility is possible if anthesis coincides in adjacent crops.

Previous cropping

An interval of 2 years should be allowed between crops of Rhodes grass (either pasture, hay or seed) unless the same cultivar is to be sown. The field should be clean, especially from other species with seeds with similar characteristics which would be difficult to separate from the Rhodes grass seed.

Difficult weeds

For the seed grower two kinds of weeds can cause difficulty: to produce clean seed all weeds or crop species with seeds similar in characteristics to those of *C. gayana* must be eliminated from the seed crop; other weeds which compete with crop growth during and after establishment must be reduced as much as possible. Starting with a clean field is essential, and pre-sowing cultivations may be aided by herbicides such as paraquat. During crop growth inter-row cultivation, particularly in the seeding year, should aim to reduce weed growth. Herbicides such as atrazine and ioxynil can be used. Hand-roguing of weed plants is usual when labour is plentiful.

Crop layout

According to Boonman (1972b), Rhodes grass yields well over a wide range of crop densities. Inter-row spacing of 25 cm proved as satisfactory as 100 cm. Low seed rates of under 1 kg/ha of pure germinating seed are satisfactory. Closer row spacing helps to produce more uniform flowering, making easier harvest.

Crop management

Chloris gayana responds well to nitrogen. In the year of sowing, nitrogen will usually be sufficient, but subsequently each seed harvest requires a dressing of about 100 kg/ha of nitrogen applied at the beginning of the next rainy season after the previous harvest. Phosphorus and potassium are not usually limiting, provided any deficiencies are corrected before sowing, although removal of large quantities of herbage after seed harvest may cause deficiency during the life of the crop, particularly of potassium. Seed growers should therefore be prepared to apply additional phosphorus or potassium if necessary. Micro nutrients such as boron, copper, manganese, zinc or molybdenum and the macro nutrient calcium are often deficient in tropical soils and any such deficiencies should be corrected to obtain good seed yields. Sulphur has also been shown to increase seed yield in some areas and may require separate application as most fertilisers are without sulphur content. The use of dung is advantageous provided it does not contain any viable seeds which might produce plants injurious in a seed crop. Seed crops are not defoliated before the first seed harvest after sowing. Removal of crop residue after harvest by grazing, cutting and carting or burning to clean up the field is usually required. There appears to be no evidence that further defoliation is beneficial, and any subsequent defoliation must be timed so that it does not remove the developing inflorescence apices as this may delay anthesis to a less favourable period. Earlier defoliation may, however, help to synchronise inflorescence development, so providing a more uniform crop at harvest. Crops sown in rows will usually fill in the inter-row spaces within a year; inter-row cultivation may be beneficial after the first (sowing year) seed harvest but is not generally possible thereafter.

Harvesting

The timing of harvest is critical. The crop will contain inflorescences at various stages of maturity. Generally, those which develop early will provide the highest yield of viable seed and, provided the weather is favourable, it is best to concentrate on these. Some shedding will, however, occur from the earliest maturing inflorescences before the crop is cut; others will have seed in the hard dough stage while the immature tillers will have seed which is still milky. Harvesting by hand can be so done that only the ripe seed is removed by beating or shaking the inflorescences against a container and it is possible to repick the field after 7–10 days. However, it must be cleaned up soon thereafter to prepare for the next seed crop. Hand-harvested inflorescences which are cut from the plant may be bound into sheaves and stooked, or if little straw is cut may be removed to a drying floor. Threshing by hand is possible or a stationary thresher or combine can be used. Swathing and picking up by combine, direct combining or using a stripper to beat the heads to shake out ripe seeds are used in areas where labour is scarce or expensive.

Seed after harvest

Drying

Hand-picked seed may be dry enough (14 per cent moisture or less) to store, but combined seed will need to be spread thinly on a well-ventilated floor and turned frequently.

Seed cleaning

The 'seed' consists of the fertile floret, the second floret (usually empty) and a further one or two rudimentary florets; the lemma is awned and usually hairy, the hairs being stiff. This presents cleaning problems as the seeds are resistant to airflow and the use of gravity separators is particularly difficult. Most efficient cleaning is achieved on an air/screen cleaner, but setting is critical and an extensive trial is required to obtain best results. Cleaned seed usually contains about 30 per cent fertile seed of which about 70–80 per cent is viable. According to Purseglove (1985) keeping the seed under good storage conditions for a year before use will improve germination as dormancy diminishes.

Diseases and pests

Thrips are reported to attack the crop in some areas.

Setaria sphacelata (Schumach.) Staph. et C. E. Hubb. Common name: setaria; South African pigeon grass; golden timothy grass

Some authors give the species name as *S. anceps* Staph. ex Massey. Three cultivars are included in OECD (1984) and both diploid and tetraploid cultivars are available. Descriptors for the cultivars have not been published.

The seed crop

Setaria is mainly cross-fertilised although weakly self-fertile. In most respects the comments made under *Chloris gayana* apply also to setaria. According to Boonman (1972a) inter-row spacing of 30–50 cm is satisfactory. Anthesis extends over a longer period and judging harvest ripeness is even more critical. The spikelets are subtended by one or more bristles which are sterile branchlets of the rachis; these remain attached to the rachis when the seed sheds. There are no awns and seed cleaning is somewhat easier than in *C. gayana*. Bunt (*Tilletia echinosperma*) is reported to occur in Kenya.

Other species

These are cross-fertilising and may be treated in a similar manner to *C. gayana*.

Panicum coloratum L.	Coloured guinea grass; Small buffalo grass
Sorghum × *almun* Parodi	Columbus grass

Sudan grass and millet

Sorgum sudanense (Piper) Staph.	Sudan grass
S. bicolor × *sudanense*	Sudan grass
Panicum milaceum L.	Common millet; proso millet

There is a total of twenty-five cultivars of these species included in OECD (1984). All are annual grasses; the sorghums are cross-fertilised and the millet partly self- and partly cross-fertilised; they thus all require isolation when grown for seed. The most difficult weed is *S. halepense* (Johnson grass) a perennial formerly sown in some areas as a forage grass but very difficult to eradicate; the seeds are only slightly smaller than those of Sudan grass. Crops are usually sown in rows up to 100 cm apart for Sudan grass, somewhat closer for millet. The seed rate is about 4 kg/ha for Sudan grass, 14 kg/ha for millet. Sowing should be delayed until the soil is warm in those areas where frosts may occur. The seed tends to ripen unevenly and harvest should be timed to obtain maximum yield from the main tillers. Swathing is preferred in those areas where the weather is favourable, otherwise the crop should be direct combined. Two cuts can be obtained from a sorghum crop in favoured areas, but the best seed yield is generally from the first.

Bracharia spp.

Bracharia decumbens Staph.	Signal grass; Surinam grass
B. humidicola (Randle) Schweichart	Koronivia grass
B. ruziziensis Germaine et C. Everardt	Signal grass; Ruzigrass
B. mutica (Forsk). Staph.	Paragrass

One cultivar of each of the first two are included in OECD (1984). Descriptors have not been published. Being apomictic the only isolation required is a gap of 3 m from crops with similar seed size. Seed crops are established in rows, and respond well to nitrogen: up to 150 kg/ha has shown good return in some areas. The seed sheds easily and is ready for harvest when it is hard, i.e. difficult to mark with the thumb-nail. Harvest is usually by direct combine with the table set low to pick up lodged panicles. In Australia suction harvesters have been used after combining to retrieve shed seed. *Bracharia mutica* is a particularly shy seed producer and is sometimes propagated vegetatively.

Other apomictic species

Cenchrus cilaris L.	Buffelgrass
Hyperrhania rufa (Nees) Staph.	Jaragua grass; Thatching grass
Melina minutiflora Beauv.	Molasses grass
Paspalum plicatulum Michx	Plicatulum
Urochloa mozambicensis (Hack) Dandy	

One cultivar of each of *C. cilaris*, and *U. mozambicensis* and two of *P. plicatulum* are included in OECD (1984). Descriptors have not been published. All are apomictic and require only 3 m isolation. They respond to nitrogen when grown in rows up to 100 cm apart. *Paspalum plicatulum* and *U. mozambicensis* are the easiest to harvest and may be direct combined. The other species are difficult as maturation is very extended, both within and between inflorescences; hand harvesting or using machines with a beating or stripping action to collect seed as it ripens are usual. The seeds of these latter species are awkward to handle, particularly *H. rufa* which has long twisted awns. *Urochloa mozambicensis* can be harvested more than once a year and in some areas up to five seed harvests have been obtained.

Partially apomictic species

Panicum maximum Jacq.	Guinea grass
Paspalum dilatatus Poiret	Dallis grass; Paspalum
Paspalum notatum Flueggc	Bahia grass
Pennisetum clandestinum Hochst. ex Chior	Kikuyu grass

There are eight cultivars of these species included in OECD (1984). Descriptors have not been published. *Panicum maximum* is apomictic but will cross-fertilise up to 5 per cent. The paspalums are also apomictic, but there are some cultivars which are cross-fertilised. Kikuyu grass is a facultative apomict. The seed grower thus has to enquire the mode of reproduction of the particular cultivar to be grown for seed before deciding upon the isolation distance required. When in doubt it is advisable to provide distances equivalent to those needed for a cross-fertilising species. These species are normally grown in rows spaced not too wide apart; they respond to grazing which tends to make inflorescence emergence more uniform. High levels of nitrogen encourage seed production. *Panicum maximum* and *Paspalum dilatatum* mature over a long period and harvest is consequently difficult; hand harvesting to remove ripe panicles is often practised. *Paspalum notatum* is easier to handle and when grazed earlier can usually be direct combined. *Pennisetum clandestinum* responds well to hard grazing, but the seed heads are then very close to the ground. Seed can be collected by cutting with a rotary mower set as close as possible (about 1 cm) to the ground; the material contains much leaf and stem and must be spread thinly to dry, or dried artificially. When dry it can be treated in a hammer-mill before cleaning with an air/screen cleaner. If the material in the field is long enough, it can be swathed and picked up with a combine. Further details on these species are given in Humphreys (1981).

7 FORAGE LEGUMES

The forage legumes cover a wide range of plant type and the different species are adapted to temperate climates with not very cold winters, to areas with hot, dry summers and often also cold winters and to tropical or subtropical conditions. Many of the species are cross-pollinating, but unlike the grasses, require insects to effect a good seed set. The flowers in many species have to be tripped, i.e. the weight of the insect alighting on the keel causes the stamens, and in some cases the stigma also, to protrude from the flower; pollination is not effective without tripping. Honey-bees are effective pollinators for many species, but for some they are not able to reach the nectar because the corolla tube is too long; in such cases humble-bees are required and the seed grower needs to ensure that the proposed area for seed production supports an adequate population of suitable pollinators.

Seed harvest is generally difficult because flowering extends over a long period and hence ripening is not synchronised. Many of the species shed seed very freely. Seed growers thus have to observe crops frequently as harvest approaches so as to judge the best time for harvest for the best seed yield.

Some species require inoculation with *Rhizobium* bacteria to achieve proper establishment. This process is discussed at the start of the next chapter (Ch. 8) on pulse crops.

SPECIES ADAPTED TO TEMPERATE CLIMATES

These species are used for pasture or hay, often in association with grasses. In some cases seed may be taken from pastures or meadows, but it is now more usual to sow crops specifically for seed production. Management is designed to gain the best possible seed yield and any forage obtained from the crop is regarded as a by-product.

Trifolium repens L. Common name: white clover

Classification of cultivars

Cultivars are grouped according to leaflet size. Generally the larger the leaflet size the shorter lived are the plants in the field, although this does not always hold true. Cultivars with large leaflets are sometimes classified as 'Ladino white clover', while those with the smallest leaflets are called 'wild white clover'. For precise measurement of leaflet size it is recommended to take the central leaflet in the third or fourth leaf from the tip of a rapidly growing stolon within 2 weeks after the mean date of flowering. Measurements are made of:

Length	Very short; Short; Medium; Long; Very long
Width	Very narrow; Narrow; Medium; Wide; Very wide

Main distinguishing characteristics of cultivars

Frequency of white leaf marks	Absent or very low; Low; Medium; High; Very high
Time of flowering	Early; Medium; Late

The seed crop

Isolation

White clover is cross-pollinated by insects, and bees are the main pollinators. For fields up to 2 ha when the seed to be produced is for further multiplication, 200 m is recommended and 100 m when the seed is for forage use. For fields over 2 ha the distances are 100 m and 50 m.

Previous cropping

Clover seed can live in the soil for several years and a period of 4 years free from white clover is recommended before sowing a seed crop. This period may be shortened if the clover is the same cultivar.

Difficult weeds

Dodder (*Cuscuta* spp.) is most objectionable and it is essential that white clover seed be free of this parasitic weed. The only sure way is to choose clean land for seed crops. Other weeds which have seeds likely to cause cleaning problems include:

Carduus sp.	Thistles
Cirsium sp.	Thistles
Cerastium vulgatum	Mouse-eared chickweed
Chenopodium album	Fathen
Geranium spp.	Cranesbill
Medicago lupulina	Trefoil
Melandrium sp.	Campion
Myosotis arvensis	Forget-me-not
Plantago major	Broad-leaved plantain
Prunella vulgaris	Self-heal
Rumex spp.	Docks and sorrels
Sherardia arvensis	Field madder
Stellaria media	Chickweed
Trifolium dubium	Suckling clover

These can be controlled by herbicides with the exception of trefoil and suckling clover.

Crop layout

It is possible to take white clover seed from a pasture or meadow, but seed yields are likely to be relatively small. For best results a crop should be sown with seed harvest in mind. There are two ways in which this can be done: a pure white clover stand managed for maximum seed yield will produce the greatest amount of seed. However, other production from the field is likely to be small and a compromise is to sow the clover with a not very aggressive grass which will provide a better pasture or forage crop when the field is not shut up for seed production. The choice depends on the relative value of the seed and the feed for livestock.

Seed crops are normally sown in the spring; they can be sown on bare ground, but it is often more economical to sow under a cover crop, generally a stiff-strawed cereal. When cover crops are used, nitrogen should be reduced from that normally given to a cereal; the objective should be to have a standing, not too heavy crop which will not smother the young clover plants.

Seed crops can also be sown in late summer, but not in the autumn as this will not permit the young plants to become established before the onset of poor growing weather. At this time cover crops should not be used.

Crops should be drilled in rows spaced about 15 cm apart. Seed rates should not be excessive as the best seed yields are obtained from uniform but not dense stands; 1½–3 kg/ha is usual in England. When a companion grass is sown the seed rate for the grass should also be reduced; 2–3 kg/ha of meadow fescue is sufficient under English conditions.

Crop management

The crop requires adequate calcium, phosphorus and potassium and any deficiency should be corrected before sowing. White clover will produce seed for several years and some of the more persistent wild white cultivars can continue to give good seed yields for 10 years or more. Some local cultivars are maintained by taking seed periodically from very old pastures. However, with most bred cultivars and for routine seed production of local cultivars, it is generally more satisfactory to plan for up to four seed harvests. After this there can be problems with weeds or other impurities which make management for seed more difficult. Crops sown in the spring will need attention in the autumn. Grazing, preferably with sheep, helps the young plants to become established and encourages them to spread into the gaps. However, grazing should not be too prolonged and poaching in wet weather must be avoided. If no stock is available the crop should be trimmed with a mower and any excessive herbage removed. Under dry conditions a rolling may be needed.

For white clover to flower freely it is necessary to avoid too much shading of the stolons where the floral primordia will develop. Therefore excessive herbage has to be removed in the spring. However, this must be carefully timed, as if done too late some of the earlier developing flowers may be destroyed and these are the ones which will give the best seed. If done too early, there may be further excessive vegetative growth which will have an undesirable shading effect on the stolons. In England, white clover will usually begin to produce buds in May or early June so that fields need to be shut up for seed by mid-May. There will be some variation between cultivars and areas in the exact timing of defoliation, but seed growers should endeavour to encourage a strong growth of early flowers. Shutting up will be later when there is strong growth in wetter seasons or areas. Grazing, preferably with sheep, is the best means of defoliation, but it is often possible to take a light cut for silage. If the herbage is not required the crop can be trimmed frequently with a mower and the cut material dispersed on the field. After grazing, fields may require trimming with a mower and rolling; if there are animal droppings they can be spread by a light harrowing before rolling.

After the first seed harvest the field will again be available for grazing or other forage use and treatment will be similar to the first year. However, there will usually be a greater amount of herbage on the field in the second or subsequent autumn which must be removed before winter sets in. If grass growth becomes excessive it can be checked by a light application of a contact herbicide such as paraquat, taking care not to apply too much as this may damage the clover. Spring vegetative growth should be removed in the same way as described for the first seed crop. Management in subsequent years is the same whenever a seed crop is desired. If no seed crop is to be taken in any one year the field can be grazed or cut for forage; however, if the latter course is adopted care should be taken not to allow excessive vegetative growth to remain too long as this can drastically reduce the clover population.

Pollination of white clover is usually by honey-bees. Most authorities suggest that one hive per hectare should be sufficient to give good pollination.

Seed-crop inspection

To check both cultivar purity and isolation an inspection at flowering time provides the best opportunity. Special care is needed in assessing cultivar purity since leaflet size can be influenced by the micro-climate in which a particular plant is growing. Some variation within a cultivar must therefore be expected.

Harvesting

The flowering period of white clover is extended and therefore seed heads ripen over a period, but the highest yield is usually obtained from the earlier formed flowers. As seed heads ripen they become brown and the seed becomes hard and is light yellow in colour. Crops should normally be harvested when about 80 per cent of the heads have changed colour. When weather is poor, harvest can begin somewhat earlier (70 per cent heads brown) as there may be danger of seed germinating in the seed head.

The greatest difficulty in harvesting white clover seed is that the crop is short and it is difficult to secure all the seed heads. Direct combining is rarely possible and the crop has to be cut and windrowed for picking up later. When crops are extremely short it is sometimes possible to cut them with a lawn-mower or forage harvester, collecting the cuttings for threshing in a stationary combine or huller. The cut material secured in this way normally contains much green matter and requires very careful handling to avoid heating and consequent loss of germination. Desiccation is not generally satisfactory in white clover, since regrowth can be very rapid and this new growth can be more difficult to deal with than the old. However, if correctly timed so that the desiccated crop can be threshed within 24 hours of treatment, this method can be successful. Applying a desiccant to an already windrowed crop can also hasten drying when there is a lot of fresh green growth on the field.

White clover is difficult to thresh satisfactorily as the seeds are tightly enclosed by the pods or 'hulls'. Cylinder clearance of combines, therefore, has to be less and speed more than for larger-seeded crops, but care should be taken to avoid damaging the seed by threshing too hard. Ordinary stationary threshers will not thresh clover seed satisfactorily and a special clover huller is needed. These machines have a double-threshing mechanism with a second or hulling cylinder; care is needed to ensure that the hulls are not left too long in the hulling cylinder as this can cause considerable damage. The hulling cylinder normally works at a speed about 15 per cent faster than the threshing cylinder. Hulling attachments can be obtained for normal stationary threshers.

Seed after harvest

Drying

Clover seed heats very quickly and germination is lost if it is stored at too high a moisture content. The threshed material often contains much green material and this must be dealt with immediately – damage can occur within a few hours. Pre-cleaning by aspiration to remove green material is worth while to reduce drying time. Natural drying by spreading on a floor in a well-ventilated shed can be achieved; the seed should not be more than 10–12 cm deep and must be turned frequently. If the floor is boarded it may be worth while to spread a hessian or similar covering. Forced air on the floor or in ventilated bins can be used, although the bulk of seed is often too small for on-floor drying to work efficiently. Air temperatures should not exceed 10 °C above ambient, with a maximum temperature of 35 °C. For continuous-flow driers the air temperature should not exceed 38 °C, and when the initial moisture content is high this should be reduced to 35 °C.

White clover seed can be stored for some months at 12 per cent moisture, but for long-term storage 9 per cent or less is required. Samples required for very-long-term storage should be dried slowly to 5 per cent moisture, sealed in suitable moisture-proof containers and stored at 0–5 °C.

Seed cleaning

An air/screen cleaner should be used in the first instance. Thereafter various specialised cleaners may be needed. For example dodder requires a velvet belt or roll cleaner or a magnetic cleaner. A gravity separator can also be useful in some instances. Cleaning white clover seed is a specialised operation requiring acquired skill.

Seed treatment

Not normally required for white clover seed.

Trifolium pratense L. Common name: red clover

Classification of cultivars

Traditionally red clover cultivars were classified as 'broad red', 'single cut' and 'late flowering' according to their earliness to start spring growth and time of flowering. Plant breeding has somewhat narrowed this range, and cultivars are now divided into two groups: early and late. The early group can be cut twice or three times in a season in northern Europe, and the late group produce most of their yield in the first cut. Within each group there are both diploid and tetraploid cultivars. Classification can thus be into four groups:

Spring growth and flowering	Early; Late
Ploidy	Diploid; Tetraploid

Main distinguishing characteristics of cultivars

Stem length at flowering time	Very short; Short; Medium; Long; Very long
Central leaflet length; upper leaf below terminal flower	Short; Medium; Long
Central leaflet width	Narrow; Medium; Broad
Time of flowering	Very early; Early; Medium; Late; Very late

The seed crop

Isolation

Red clover is cross-pollinated, largely by bees. Isolation is required as described for white clover. Diploid cultivars will not set seed if pollinated by

Fig. 7.1 A well-isolated red clover seed crop

tetraploid or vice versa; isolation of at least 50 m is required to avoid excessive seed yield reduction (Fig. 7.1).

Previous cropping

Red clover seed can live in the soil for long periods and is also prone to attack by two soil-borne diseases: clover rot (*Sclerotinia* sp.) and eelworm (*Ditylenchus* sp.). To avoid both problems requires an interval free from red clover of at least 6 years. When there is no danger of soil-borne infection and the seed to be produced is to sow forage crops, this interval may be shortened to 2 years.

Difficult weeds

Dodder (*Cuscuta* spp.) is a major problem in some areas and very difficult to control; it is essential to choose land free from dodder for red clover seed production. Other weeds which may cause difficulty for producing a clean seed sample are:

Cardenuus spp.	Thistles
Cirsium spp.	Thistles
Geranium spp.	Cranesbill
Medicago lupulina	Trefoil
Melandrium sp.	Campion
Plantago lanceolata	Ribgrass
Rumex spp.	Docks and sorrels

With the exception of trefoil, these weeds can be controlled by herbicides.

Crop layout

Early-flowering red clover is usually sown pure for seed production. Late-flowering cultivars give the best results when sown without a companion, but in some areas a grass is added; however, aggressive companion grasses can reduce seed yield so it is best to choose a species or cultivar which is not too vigorous. It will be an advantage if any companion grass will produce seed at the same time as the late-flowering clover; timothy has been used in England.

Red clover seed crops are normally sown in spring under a cover crop such as a cereal. Management of the cover crop should aim to avoid too heavy a cover which would depress the development of the young clover plants. Later sowing without a cover crop is possible, but must be in time for healthy, vigorous clover plants to develop before growth is reduced in the winter. In some areas it is possible to take a seed crop in the year of sowing by sowing early in the season without a cover crop. However, late-flowering cultivars from the higher latitudes are adapted to flower under long-day conditions and if sown in lower latitudes may produce a poor yield of seed from the earlier-flowering plants only; this treatment can change the characteristics of the cultivar. There is less danger of such a shift in early-flowering cultivars and seed yields are likely to be more satisfactory.

The seed may be broadcast, but in most areas will benefit from sowing in rows spaced 15 cm apart. Some late-flowering cultivars may do better in wider-spaced rows up to 60 cm apart provided lodging is not excessive. For the closer spacing about 5 kg/ha seed of diploid cultivars is required with

1 kg or less when the rows are wide apart. These quantities are doubled for tetraploids.

Crop management

When seed crops of early-flowering cultivars are sown early in the season and taken for seed in the same season, the first growth is allowed to grow unchecked and seed harvested at the first opportunity. When sowing is in one season with the intention of taking seed the next grazing in the autumn is usually beneficial. Fields should in any case be cleared of excess herbage by trimming at the end of active growth so as to leave uniform plants. When cover crops are used they must be removed in good time so as to allow the seedling clover plants to develop.

Early-flowering cultivars are usually taken for seed at the second cut in the year after sowing. High seed yields can be obtained from the first cut, but foliage growth is generally excessive with severe lodging and harvest extremely difficult. The first cut for hay or silage should be timed so as to allow the second growth to flower and set seed at a time when harvesting conditions are likely to be good. In England the first cut is normally in late May or early June.

Late-flowering cultivars are best taken for seed at the first growth of the season as lodging is less likely. Some intermediate cultivars may benefit from a light early grazing or cutting to remove excess foliage, but this must not be left too late.

Early-flowering cultivars are normally harvested for seed only once from each crop. Late-flowering cultivars may be kept for a second year but seed yields are normally less than in the first. A third seed harvest is sometimes possible in favoured areas but crops will not stand for longer than this.

Pollination of red clover is mainly by bees. The length of a corolla tube and height of the nectar in the tube govern the effectiveness of pollinators. For early-flowering cultivars the honey-bee is generally effective, but its tongue is too short for it to reach the nectar in the longer corolla tubes of late-flowering cultivars. The humble-bee *Bombus terrestris* will take nectar by cutting holes in the base of the corolla and these holes are also used by honey-bees; this behaviour avoids tripping the flower and hence pollination. Other species of humble-bee, for example, *B. hortorum, B. ruderatus* or *B. subterraneus* have longer tongues, do not rob and are effective pollinators. If honey-bees are used, about one hive per 3 ha is required to provide sufficient pollinators. When relying on humble-bees it is important to provide suitable nesting sites in rough, unweeded areas and not to sow too large an area of seed crop; the humble-bee population generally pollinates more effectively in fields of less than 5 ha. Attempts have been made to domesticate pollinator humble-bees by importing queens to the vicinity of a seed crop; this has been successful in some instances.

Seed-crop inspection

Cultivar characteristics can best be seen when the seed crop is beginning to flower and it is intended to take seed from that growth. Isolation can also be checked at the same time.

Harvesting

Since flowering may extend over some weeks, judgement of when to harvest can be difficult. However, in general the earlier flowers produce the best seed and growers should concentrate on these. The seed is ready for harvest when the majority of the heads have changed colour to brown; the seed can be rubbed out in the hand and will be dark coloured and hard. The proportion of fully ripe heads will be between 60 and 90 per cent depending upon cultivar and local conditions.

Direct combining is often possible and is to be preferred. However, the amount of green material is often too great and it is then necessary either to windrow the crop before picking up or to desiccate with diquat or other desiccant before direct combining. The latter course is only possible when a few days of dry weather can be foreseen as regrowth can occur in showery weather. Desiccation should take place when the crop is fully ripe, whereas windrowing can be done some days before this stage is reached because seed will mature further in the windrow.

Red clover seed flows easily and combines and other equipment should be sealed to prevent seed loss. The crop threshes more easily than white clover, but can be difficult and careful setting of the combine is required. Cylinder clearance is somewhat greater than for white clover. If the crop is to be threshed in a stationary thresher it is preferable to use a clover huller or to add a hulling attachment to an ordinary thresher. Care should be taken not to thresh too hard as injury to the seed can result.

Seed after harvest

Drying and seed cleaning

The comments made about white clover seed apply to red clover. However, the quantities to be handled are usually greater and bulk drying facilities are therefore more easily adapted for use.

Seed treatment

Chemical dressing against fungus diseases is not generally used for clover seed. The stem eelworm (*Ditylenchus* sp.) is seed borne and seed should not be harvested from infected crops. If infected seed is detected it should be fumigated. This is a specialist operation and a qualified service should be employed.

Other temperate clovers

Trifolium hybridum L.

Alsike is a short-lived perennial and it is possible to take four seed harvests from one sowing. Both diploid and tetraploid cultivars are available. Crop establishment and management are similar to that required for red clover. Pollination is by honey-bees. Seed is normally taken from the first growth without any spring defoliation. Harvesting is similar to that described for red clover. Some autumn grazing is beneficial and excess herbage should be removed before growth ceases for the winter.

Trifolium fragiferum L.

Strawberry clover is a perennial which resembles white clover in many respects but is much more tolerant of extremes of temperature, being able to withstand high summer and low winter temperatures. Several seed crops can be taken from one sowing, but the crop will not set seed before the first winter. Dense crops are said to be best for seed production and it withstands hard grazing well, but should not be defoliated too closely before winter as this encourages winter-kill. Strawberry clover is normally regarded as cross-pollinating, but self-fertile forms exist; seed growers should therefore seek advice from plant breeders before deciding on the isolation distance required. Crop establishment and management are similar to that required for white clover. Seed is ready for harvest when the majority of the capsules are light brown; the capsules readily break off and can be blown away if harvest is delayed too long. Handling in damp weather – for example during early morning dew – may prevent some seed loss.

Trifolium alexandrinum L.

Berseem or Egyptian clover is an annual. In some areas it can be sown in the autumn for seed harvest in the following year. It is self-fertilised, but requires visitation by insects to trip the flowers to achieve good seed set. Seed is normally harvested from the first growth after sowing. The crop can be combined direct.

Trifolium incarnatum L.

Crimson clover is an annual. Fertilisation is similar to berseem and it can be sown in autumn in some areas. Seed yields are high, but there is little or no dormancy and seed germinates very quickly. The crop is difficult to combine direct because the stems remain green when the seed is ripe. It is therefore usually necessary to windrow when 75 per cent of the seeds are moderately hard. Windrows should be kept small to ensure an even feed when picking up with the combine. The seeds are difficult to hull and careful combine setting is needed. In many instances subsequent hulling in the seed-cleaning plant will be needed.

Trifolium resupinatum L.

Persian clover is an annual which can be sown in autumn in some areas. It is self-fertilised and insects are not normally required for seed production. Some grazing is possible in the spring to reduce vegetative growth at harvest, but should cease at least 1 month before flowering; this defoliation helps to produce more uniform flowering and hence seed ripening. The seed is ready for harvest when the majority of capsules have turned light brown. The capsules are balloon-like and are easily dispersed by wind, therefore wind-rowing is preferable to direct combining.

Trifolium semipilosum Fres.

Kenya clover is a perennial similar to white clover. It has a very specialised rhizobium requirement and therefore requires inoculation on most soils.

Trifolium vesiculosum Savi.

Arrow leaf clover is an annual which is normally autumn sown for harvest the following year. It is pollinated by bees.

Medicago sativa L. and *Medicago* × varia Martyn. Common name: lucerne; alfalfa

Classification of cultivars

Cultivars are classified as Early; Medium; or Late. Early cultivars start growth early in spring, flower early and are generally erect in growth habit. Late cultivars start growth about 4 weeks later, flower 4 weeks later and are not so erect in growth habit. There are also a few cultivars which have a more or less creeping habit with rhizomes.

Main distinguishing characteristics of cultivars

Length of central leaflet on third or fourth leaf below inflorescence	Short; Medium; Long
Width of same leaflet	Narrow; Medium; Broad
Stem length	Very short; Short; Medium; Long; Very long
Dominant flower colour	White; Yellow; Light blue-violet; Dark blue-violet; Red-violet
Time of flowering	Very early; Early; Medium; Late; Very late

The seed crop

Isolation

Lucerne is pollinated by certain kinds of bees which are able to 'trip' the flowers. Recommended isolation distances are 200 m for fields of 2 ha or less where the seed is intended for further multiplication, and 100 m for fields over 2 ha. When the seed is intended for forage crops the distances are 100 m and 50 m.

Previous cropping

Lucerne seed can live in the soil for several years. An interval free from lucerne of 4 years is recommended before sowing a seed crop. In the earlier stages of a multiplication cycle a longer interval, up to 6 years, is advised. These intervals can be shortened if authentic seed of the same cultivar was used to sow the preceding crops.

Difficult weeds

Dodder (*Cuscuta* spp.) is very difficult to control in lucerne seed fields and the seed cannot always be separated after harvest; fields free of this weed should be chosen for seed crops. Other weeds which have seeds about the same size as lucerne include:

Amaranthus spp.	Pigweed
Brassica spp.	Mustard, rape
Cenchrus spp.	Sandbur
Convolvulus arvensis	Bindweed
Galium aparene	Cleavers
Melilotus spp.	Sweet clover
Rumex spp.	Docks and sorrels
Sida spp.	Mallow
Sorghum halepense	Johnson grass

Crop layout

Higher seed yields are obtained from crops sown in rows rather than broadcast. The appropriate distance between rows depends upon the growing conditions. In general, conditions which encourage vigorous growth and large plants require a wider spacing than those which restrict growth; an exception to this is in very dry areas where a wider spacing will normally provide enough plants for the available moisture. In England, 35–60 m between rows constitutes the optimum range. In the USA, some conditions require spacing of 150 cm or more.

Lucerne is best sown without any companion grass to achieve good seed yields. Seed rate should not be excessive, the aim being to achieve a uniform stand of plants in the row with about 10 cm between plants. In good conditions this can be achieved with 1 kg/ha when the rows are 90 cm apart, or 3 kg/ha for 35 cm between rows. Rates should be increased when conditions are less than optimum.

In Europe, lucerne seed crops are usually sown in spring, under a light cereal crop. In the USA, autumn sowing is possible in some areas and will give a satisfactory seed yield in the following year. It is also possible to obtain a good seed yield by sowing early in the season and taking the first growth for seed in the same season.

Crop management

Once established, lucerne seed crops can continue to produce seed for several years, but usually about 5 years is the most that can be expected to give good seed yields. The field should be well supplied with calcium, phosphorus and potassium and any deficiencies should be made good before sowing. Lucerne is deep rooted and relies on nutrient supplies at a greater depth than many other crops.

Crops should be trimmed before winter sets in, without cutting too close to the ground; the objective is to remove excess herbage and have a uniform stand. In spring, the first growth should be allowed to flower and set seed. In some areas it may be possible to take a forage cut and then seed from the second growth, but usually this means a late autumn, early winter harvest when weather is less reliable. Seed from the first growth can usually be harvested in time to allow a late autumn forage cut to be taken.

Management in subsequent years follows the same pattern: the crop is cut in the late autumn to provide a uniform stand into the winter, and seed is taken from the first growth in the following year.

Weed control is important as lucerne is a poor competitor in the seedling stages. Measures appropriate for forage crops are equally appropriate for seed crops. Lucerne is attacked by numerous insects and these should be controlled by the application of suitable insecticides; some insects attack the flowers and seeds and these may require special control measures not usually required in forage crops. Examples are *Lygus* bugs, chalcids and crickets. Control measures must take account of the need to encourage pollination by bees, and therefore spraying insecticides at the time of flowering should be avoided, or if it must be done, should take place at a time of day when bees are not active. In some areas flowering can be timed by taking appropriate forage cuts to avoid the most damaging insect attacks.

Pollination is by bees. Honey-bees are the most important since they are readily available, and in most instances it is possible to set hives in the crop. However, the flowers are not 'tripped' by bees collecting nectar to anything like the extent to which tripping takes place when bees are collecting pollen. According to Hanson (1975) pollen collectors are forty-five times more efficient pollinators than nectar collectors. It is not known what causes honey-bees to collect pollen in preference to nectar. To be effective Hanson recommends a density of one bee per square yard (0.84 m^2) when collecting pollen but up to ten when collecting nectar.

Alternatives to honey-bees have been sought in some areas. In temperate areas the humble-bee, *Bombus terrestris*, can pollinate lucerne. In hotter areas the alkali bee, a ground-nesting species, is effective. The bees nest in moist loams on a bare level site (Fig. 7.2(a)). They do not like rain, and thus only succeed in warm, dry areas. It is possible to provide artificial nesting sites and to encourage bee colonies to develop, but in recent years more attention has been given to the leaf-cutter bee.

Leaf cutters are easier to manage than alkali bees. They nest in small holes in wood or similar situations. Artificial nesting sites are provided by erecting shelters in which is placed wood drilled to a depth up to 15 cm (see Fig. 7.2(b)); alternatively, drinking straws can be used or any material which has small-diameter holes in it. Plastic materials are less suitable in humid areas, and holes should be shallower in such areas to ensure adequate ventilation. Colonies of leaf cutters can be purchased in the USA and are relatively easily established in artificial nests sited at strategic intervals in a lucerne seed field. The bee colonies can be stored over winter, and by management of the storage temperature the colonies can be encouraged to emerge when the lucerne flowers begin to open. The bees remain in the nests at temperatures below 20 °C and emerge when temperatures rise above 30 °C. They are effective only in areas where day temperatures are above 30 °C. Colonies are required to provide a density of 1235 female bees per hectare, as it is the females which do most of the foraging and hence are the most effective pollinators. Some beetles and chalcid wasps are predators of the leaf cutters; these are best controlled during the storage period. Cells can be removed from the nesting holes and the predators cleaned out. The cells can be stored, but are more vulnerable and need careful handling. In spring the emerging bees are encouraged to return to the nesting material by placing them in trays in the shelters, covered with sawdust to prevent drying out.

(a)
Fig. 7.2 (a) Alkali bees' nesting site; (b) leaf cutter bees

Seed-crop inspection

To identify off-types, seed crops should be inspected when in full flower. At this stage isolation can also be checked.

Harvesting

Lucerne plants flower over a considerable period and therefore the seed does not ripen all at the same time. Seed growers have to decide when to harvest so as obtain the highest yield of good seed. The seed is in small, spirally twisted pods and the best time for harvest is when three-quarters of the pods have changed colour to brown or dark brown, and before many begin to dehisce.

Because ripening is extended it is usually not possible to combine seed crops directly without other treatment. The alternatives are to windrow or to desiccate. The more uneven the flowering and therefore ripening, the more likely it is that windrowing will be the best course. The crop can be cut earlier and some immature pods will ripen after cutting. Desiccants such as diquat have the advantage that seed can be allowed to mature to the optimum harvest stage before desiccation, and can then be direct combined. However, the disadvantage is that if combining is delayed more than 2–3 days after desiccation the crop may shed seed or secondary growth may occur.

Combining is generally possible when the leaves are dry – that is, below 20 per cent moisture. At this stage stems may still be green and will have a higher moisture content, especially in desiccated crops.

Grain combines are satisfactory for lucerne but may need to be sealed to prevent seed loss as the small seed flows freely. Careful setting is required to avoid damage to the seed and to prevent excessive seed losses.

(b)

(b)

Fig. 7.2 *(cont.)*

After harvest, straw and chaff should be removed from the field. It may contain harmful insects and should be destroyed in areas where these are prevalent. Burning *in situ* is effective and will not harm the lucerne plants.

Seed after harvest

Drying

Seed should be below 12 per cent moisture content for medium-term storage, and for longer periods over 1 year 8 per cent or less is required. The temperature of the drying air should not exceed 38 °C when initial moisture content is below 20 per cent; if higher, air temperature should be reduced below 30 °C.

Seed cleaning

A gravity separator and air/screen cleaner are normally required. Roll separators and magnetic separators are also often needed.

Seed treatment

Inoculation with *Rhizobia* is often required. Other seed treatments are not generally necessary. Seed infected with stem eelworm (*Ditylenchus* sp.) should be fumigated using a specialised fumigation service.

Hybrid cultivars

Lucerne cultivars are synthetic in that they represent the controlled multiplication from selected parent plants. Cytoplasmic male sterility has been found and hybrid cultivar production is possible. However, problems with low seed yield have so far prevented the widespread use of hybrids.

Onobrychis viciifolia Scop. Common name: sainfoin

Thirteen cultivars are listed by OECD (1984). There are two main types of sainfoin: a short-lived biennial which survives for no more than 3 years, and a longer-lived perennial. The crop requires soil with a high calcium content. It is slow to establish, but once established resists drought well. Sainfoin is self-incompatible and therefore cross-fertilised; seed crops require good isolation. The main pollinating agents are bees, and from two to ten hives per hectare have been recommended. To achieve a good seed set it is estimated that each flower should be visited by a bee on at least three occasions. The seed is usually under-sown in a cereal nurse crop in rows 15–20 cm apart. The seed rate is about 50 kg/ha if the seed has been hulled and about 80 if unhulled. The crop should not be cut or grazed after removal of the nurse crop unless there is excessive growth to remove before winter. The biennial form is first cut for hay in the second year and seed is harvested from the second growth. The perennial form cannot be treated in this way, and seed must be harvested from the first growth. Best seed yields from the perennial form are usually in the second or third harvest years, and crops are sometimes taken for hay only in the first year; these crops can continue for 7 or more years. Harvesting, drying and seed cleaning are similar to that described for red clover.

Ornithopus sativus Brot. Common name: serradella

Serradella is mostly grown in the coastal areas of Spain, Portugal and Morocco and is of only minor importance elsewhere. It is largely self-fertilised and therefore requires only sufficient isolation to avoid mixtures at harvest. For seed it is normally treated as an annual and has been grown with a companion crop of spring rye to provide support. Seed rate, using unhulled seed, is between 20 and 40 kg/ha, depending on conditions. The seed sheds easily and harvest is usually by windrowing and subsequent picking up by combine. Serradella is susceptible to attack by the fungus *Colletotrichum trifolii* and requires seed treatment. Inoculation with rhizobium is desirable in areas where the crop has not been grown before.

Vicia spp. Common name: vetches

There are three species of *Vicia* which are commonly grown for seed as forage crops. They are (with the number of cultivars listed by OECD, 1984 in parenthesis):

V. sativa L.	Common vetch (41)
V. villosa Roth.	Hairy vetch (16)
V. pannonica Crantz.	Hungarian vetch (2)

Classification of cultivars

The vetches are annuals grown in temperate areas. Some cultivars, especially of *V. villosa*, can be sown in the autumn and are winter hardy in areas where the winter is not too severe. Descriptors have been published by OECD (1973) and by UPOV (1976) for *V. sativa* only: A first classification can be made on seedlings grown in a glasshouse:

Number of branches 3 to 4 weeks after emergence	Absent or very few; Few; Medium; Many; Very many

Main distinguishing characteristics of cultivars

Shape of first seedling leaflets (ratio: width/length)	Very narrow; Narrow; Medium; Wide; Very wide
Anthocyanin coloration of leaf axil	Absent or very weak; Weak; Medium; Strong; Very strong
General shape of leaflet	Round; Long
Shape of tip of leaflet	Highly convex; Convex; Straight; Concave; Deeply concave
Anthocyanin coloration of nectaries	Absent; Present
If present, intensity of coloration	Very weak; Weak; Medium; Strong; Very strong
Colour of standard petal	White; Pink; Light violet; Dark violet
Hairiness of pod	Absent or very weak; Weak; Medium; Strong; Very strong

Constrictions between seeds in pod	Absent or very weak; Weak; Medium; Strong; Very strong
Seed shape	Spherical; Subspherical; Sublenticular; Pulvanate; Rectangular solid
Seed size	Very small; Small; Medium; Large; Very large
Ground colour of seed testa	White; Grey-brown; Grey-green; Blue-black
Main ornaments on seed testa	Absent; Present
Type of main ornaments on seed testa	Punctuation; Speckling; Mottling; Marbling
Second ornaments on seed testa	Absent; Present
Type of second ornaments on seed testa	Punctuation; Speckling; Mottling; Marbling
Colour of main ornaments on seed testa	Brown; Violet; Dark grey; Sepia
Colour of cotyledons	*Café au lait*; Pink-violet; Orange
Time of 10 per cent flowering	Very early; Early; Medium; Late; Very late

The seed crop

Isolation

V. sativa is almost entirely self-fertilised while *V. villosa* is largely cross-fertilised. Isolation requirements thus differ from 3 m to that required for cross-fertilised crops, normally 200 m and 100 m for crops 2 ha or less when the seed to be produced is for further multiplication or to sow crops for fodder production; and half these distances for larger fields.

Previous cropping

Vetches produce a high proportion of hard seed which can live in the soil for many years. When the crop is to produce seed for further multiplication a 6-year interval free from vetch crops should be allowed before sowing. When the seed is intended for the production of fodder crops a shorter interval is possible. Shorter intervals are possible when special measures are taken to eliminate volunteer plants or when the same cultivar is to be grown again.

Difficult weeds

Docks and sorrels (*Rumex* sp.) and some clovers can cause difficulty during seed cleaning.

Crop layout

For seed production vetch should be sown without any companion crop in rows spaced up to 20 cm apart. Seed rate is between 20 and 40 kg/ha.

Crop management

A seed crop does not require management different from a fodder crop except that the first growth is allowed to mature seed before harvest. Bees visit vetch flowers, and for *V. villosa* in particular it is worth while to introduce one or two hives per hectare. Seed-crop inspection should take place when the crop is in flower.

Harvesting
A severe set-back by, for example, drought during flowering may cause one group of flowers not to set seed. Crops should therefore be examined prior to harvest to ensure that pods are properly filled. The two most commonly grown species *V. sativa* and *V. villosa,* shed seed very easily, and it is necessary to time harvest so as to secure seed from the best pods: these will normally be the lower pods which ripen first. Most crops are now direct combined when the seed in the lower pods is firm and has changed colour; at this time the upper pods will still be green and the seed soft. As shedding is quick, it is necessary to have enough combine strength to complete harvest by the method in 3–4 days. An alternative is to windrow and pick up with the combine after about a week; however, seed losses by this method are likely to be higher. *Vicia pannonica* does not shed seed so readily and may be left until about 90 per cent of the seed is ripe; however, it is more difficult to thresh and careful combine setting is needed to avoid leaving seed in the pods.

Seed after harvest

Drying and seed cleaning
Seed from the combine will usually contain green material and will require drying as described for white clover; vetch seed is larger than white clover and therefore somewhat more easily handled. Seed cleaning can be achieved on an efficient air/screen cleaner. Seed treatment is not required, but seed will often contain larvae of the vetch weevil (*Bruchus branchialis*); such seed should be fumigated in store although as infected seed is lighter it can sometimes be removed during cleaning.

SPECIES ADAPTED TO AREAS WITH RELATIVELY HOT SUMMERS

This group of species are generally grown in North America and the Mediterranean area and may be adapted to particular areas. Some are used mainly for soil conservation.

Lespedeza stipulacea Maxim.
Common name: Korean lespedeza

Two cultivars are included in OECD (1984), but no detailed descriptors have been published by the international organisations. It is an annual grown extensively in the south-east of the USA.

The seed crop

Isolation
There are two types of flower: one does not open and is self-fertilised, while

the other can be self-fertilised but is also visited by bees – both honey-bees and humble-bees. Putting two hives per hectare, honey-bees in the seed crop is said to improve seed yield. Isolation as for a cross-fertilising crop is desirable.

Previous cropping
An interval of 4 years free from lespedeza is required before sowing a seed crop, unless the same variety is to be sown again. Seed in store, however, deteriorates quickly and should not be kept more than 2 years.

Difficult weeds
Dodder (*Cuscuta* sp.) can be a serious weed in lespedeza crops and clean land is essential for seed production. Other common weeds are Johnson grass (*Sorghum halapense*) and ragweed. (Ambrosia spp.)

Crop layout
The crop is usually drilled and if unhulled seed is used the seed rate is around 25 kg/ha.

Crop management
If growth is excessive it can be removed as a hay crop and the second growth is then taken for seed. However, if any of the leaves have begun to drop from the lower part of the stem the crop should not be defoliated before taking seed as many plants may be killed if there are no live buds below the cutting point. Lespedeza crops often continue for some years because fallen seed re-establishes the plants. However, this practice is not favoured when seed of a specific cultivar is to be produced. Seed-crop inspection should take place at flowering time.

Harvesting
Lespedeza sheds seed quite readily when it is mature and harvest should not therefore be unduly delayed. The crop is usually combined direct. For combining very weedy crops the cylinder speed will have to be increased.

Seed after harvest

Drying
In those areas where lespedeza is grown the seed will usually be harvested at a low enough moisture content for storage. However, if there is much green material with the seed it will need to be dried immediately. Pre-cleaning such seed is usually worth while to reduce drying time.

Seed cleaning
This can usually be achieved on an air/screen cleaner, but if dodder is present a velvet roll mill will be needed. The seed is usually left unhulled as the dormancy period is relatively short.

Other species of Lespedeza

L. striata (Thunb.) Hook et Corn	Striata lespedeza
L. cuneata (Dum.) G. Don	Sericea lespedeza

Lotus corniculatus L. Common name: bird's-foot trefoil

This is an important perennial species as a forage plant on poorer soils in the temperate areas. There are eighteen cultivars listed by OECD (1984) but no descriptors have been published internationally.

The seed crop

Isolation

Bird's-foot trefoil is self-incompatible and therefore cross-fertilised, generally by bees. It requires adequate isolation: for small fields up to 2 ha, 200 m when the seed to be produced is for further multiplication, and 100 m when it is intended for sowing forage crops; for larger fields over 2 ha the distances are 100 m and 50 m.

Previous cropping

Trefoil seed can remain viable in the soil for long periods. When a seed crop is for further multiplication an interval of 5 years free from trefoil should be allowed prior to sowing; for crops to produce seed for sowing forage crops, 2–3 years is usually considered adequate. These intervals may be shortened when the same cultivar is to be sown again, or when special measures are taken to control volunteer seedlings.

Difficult weeds

Bird's-foot trefoil has a relatively open growth habit and so is subject to weed competition. Most can be controlled by herbicides. Weeds which have seeds difficult to remove from trefoil include bindweed (*Convolvulus arvensis*) and docks and sorrels (*Rumex* sp.). Other forage legumes – clovers and lucerne – are also liable to cause difficulty in this respect.

Crop layout

A uniform stand gives the best seed yields and therefore drilling is generally preferred to broadcasting. Rows 15–60 cm apart can be used, but the closer spacing tends to give plants with a shorter range of flowering time so that seed ripening is more uniform. Seed rates are about 3–5 kg/ha. In some areas inoculation with rhizobium is required. Companion grasses may enhance the forage value of the crop, but generally reduce seed yield and are not therefore recommended in a seed crop.

Crop management

Bird's-foot trefoil is perennial and several seed harvests can be obtained from a crop. Normally 5 years are considered to be the limit, but crops can continue for much longer provided the build-up of volunteer plants is not excessive. Establishment after sowing is slow and for this reason crops are usually sown in spring, especially in colder areas; later sowing is possible in favoured areas and where irrigation is available. The first seed harvest is taken in the year after establishment. Crops should be trimmed of excess herbage to leave a uniform stand into the winter. In the following spring the first growth is the best for seed. However, if it is desired to take the seed harvest at a later, more favourable, time a forage cut or grazing can be taken; from defoliation to harvest takes about 2 months, but as flowering is also controlled by day-length it should not be left too late. After the first harvest the seed crop is managed in much the same way in subsequent years, alternating judicious grazing or cutting with a seed harvest at the best time of year. Trefoil does not generally require high fertility, and crops which produce excessive foliage on rank growth do not produce much seed. There should not, however, be deficiencies of phosphorus or potassium although the crop does quite well under acid conditions. Bees are the best pollinators and two hives per hectare are recommended. Seed-crop inspection is usually recommended at flowering time when it is also possible to check isolation.

Harvesting

Bird's-foot trefoil is one of the more difficult legumes to harvest for seed. The pods readily dehisce when ripe, and ripening is extended over a considerable range, although some cultivars have been selected for more uniform ripening. Windrowing is preferred to direct combining. Crops should be watched closely as harvest approaches. The plants remain green as the seed ripens; pods turn dark brown to black and the seed becomes firm, being yellow to brown in colour when ripe. Defoliants (e.g. diquat) can be used and help to reduce the amount of green material to be harvested; however, regrowth can be quite rapid so that harvest should follow within 7 days of application. Windrowing and picking up when the crop is damp with dew helps to reduce shedding losses.

Seed after harvest

Drying and seed cleaning

As the stems and leaves are usually green when the seed is threshed there will be green material in the seed which must be dried immediately to prevent heating. Requirements are similar to those described for white clover.

Seed treatment

Hard seed content is generally high and scarification of the seed before sowing is sometimes beneficial. Inoculation with rhizobium is desirable in some areas. Treatment with insecticides or fungicides is not considered to be worth while.

Other species of Lotus

Numbers of cultivars listed by OECD (1984) in parenthesis.

L. tenuis Waldst. et Kit ex Willd	Slender bird's-foot trefoil (1)
L. uliginosus Schk.	Greater bird's-foot trefoil (1)

Slender bird's-foot trefoil is very similar to *L. corniculatus*. Greater bird's-foot trefoil is somewhat different and requires a different strain of *Rhizobium*. The seed is yellowish-green when ripe although some other colours may also occur.

Melilotus spp. Common name: sweet clover

There are two species of *Melilotus* which are grown for seed:

M. alba Med.	White sweet clover
M. officinalis (L.) Pall.	Yellow sweet clover

The OECD (1984) lists two cultivars of the former and three of the latter. No descriptors have been published. There are both annual and biennial forms in the two species, and there is a wide range of maturity times between cultivars which can differ in time of ripening by up to 2 months.

The seed crop

Isolation
Although some self-fertilisation occurs the flowers have to be tripped by insects, usually bees. Thus a rather high degree of cross-fertilisation occurs, although this is said to be more in white sweet clover than in the yellow. Isolation as for a cross-fertilised crop is therefore required (see under *Lotus corniculatus*). Two to twenty-five hives of bees per hectare have been used.

Previous cropping
There is a high proportion of hard seed so that seed can remain viable in soil for a long period. For crops growing seed for further multiplication a 5-year interval free of sweet clover is required, and for other seed crops two or three years.

Difficult weeds
Sweet clover grows vigorously when established and will compete well with weeds. Weed control before sowing and during establishment will normally achieve a clean crop.

Crop layout
Best seed yields are obtained from stands which are not too thick as very rank tall growth prevents full pollination and is much more difficult to harvest. Drilling is preferred to broadcasting and the rows may be 15–100 cm apart, the wider spacing being favoured in drier areas. At close spacing up to 15 kg/ha will be required; this is reduced to 3 or 4 kg in the very wide rows. Companion grasses can be used with sweet clover and may improve the fodder value of any cut or grazed herbage.

Crop management
Seed crops are sown in the spring for harvest in the following year. Best seed yields are obtained from crops which are grazed in the establishment year. Crops should be shut up for winter with about 15 cm growth which has been trimmed uniformly. In the harvest year the highest potential seed yield is from the first growth. However, this can grow to 120 cm high and create difficult harvesting problems, and it often pays to graze early or to take an early, light hay cut to reduce the amount of vegetation at harvest; if cut for hay too late, seed yields can be seriously reduced. Seed-crop inspection is most effective when the crop is in full flower.

Harvesting
Sweet clover sheds seed very quickly and therefore windrow harvesting is preferred to direct combining; this also allows the herbage to dry before threshing. The crop is ready for windrowing when about 60 per cent of the pods have changed colour to brown or black; the stems at this stage are still sappy, and only a few of the leaves will have fallen. Desiccation with a chemical such as diquat followed by direct combining is possible, but combining must be done within 3 days of applying the chemical, otherwise harvesting becomes even more difficult. Seed shedding is least when the crop is damp and early morning harvesting is better than mid-afternoon. In very dry areas, harvest can begin before the 60 per cent colour change stage is reached.

Seed after harvest

Drying, seed cleaning and seed treatment
Seed from the field will usually contain green material and should be carefully dried immediately. The combine will leave a high proportion of unhulled seed as the hulls are quite difficult to remove. A scarifier will remove hulls effectively and will also reduce hard seed content. Seed treatment is not normally necessary.

Coronilla varia L. Common name: Crown vetch

Three cultivars are listed by OECD (1984) but no international descriptors have been published. The principal use for this species is in soil conservation. It grows well and persists under adverse conditions, being perennial. The flowering stems reach a height of 90 cm and the herbage below forms a dense mat some 30 cm thick. Crown vetch is cross-fertilised by humble-bees, but honey-bees are also effective. Adequate isolation is required. The seed sheds easily and direct combining is the best harvesting method. Harvested seed requires to be dehulled in a scarifier, but must be handled with care as the seed coat is easily damaged. Scarification to reduce hard seed content must also be done carefully.

Aestragalus cicer L. Common name: chick-pea milk vetch, cicer milkvetch

One cultivar is listed by OECD (1984). Chickpea milk vetch is very slow to establish and requires careful seed-bed preparation. Seed crops should be sown in rows 60–90 cm apart at a seed rate of 3–6 kg/ha. The crop is cross-fertilised; the main pollinating agents are humble-bees, but honey-bees are also effective. Adequate isolation is required. Chickpea milk vetch does not shed seed, but is difficult to thresh. Harvest is by windrowing and picking up by combine; however, the windrow must be allowed to dry thoroughly and the combine should be run with a high cylinder speed. It is sometimes economic to pick up and thresh the crop a second time.

Medicago lupulina L. Common name: black medic trefoil

One cultivar is listed by OECD (1984). The species is a biennial and is usually sown one year for seed production in the following year. Grazing will increase stem production, but should not continue late in the spring. The plant is low growing and may only reach 8–10 cm, but is more normally up to 40 cm tall. The plant flowers first on the lower parts of the stem and these seeds may shed before the majority of the pods are ready for harvest. Ripe pods turn black. The seed is small and difficult to handle. The crop is usually windrowed and picked up within one or 2 days; the optimum harvest period is short and may last no more than a week. The seed is closely enclosed in the hulls and will require scarification to remove them before cleaning.

SPECIES ADAPTED TO VERY DRY CONDITIONS

Most of the work on these species has been done in South and Western Australia, but they are also adapted to large areas in South America and elsewhere. They are characteristically difficult to harvest because the seed is deposited on or under the ground as it ripens. Once sown in a field, the crop reseeds itself each year, and as the seed varies in dormancy period it is not possible to identify the generations. The OECD has therefore prepared a separate scheme for these species (OECD 1977). All the cultivars are self-pollinating and are identifiable from marker genes. Basic seed is produced by the plant breeder and thereafter only certified seed is produced, which may be from fields originally sown with basic seed, but subsequently reseeded naturally. A field may continue to produce certified seed indefinitely, provided there are no more than 5 per cent of plants which are not characteristic of the cultivar.

Trifolium subterraneum L. Common name: subterranean clover

Main distinguishing characteristics of cultivars

Characteristics of cultivars were published by OECD (1973):

Pattern of pale central mark on leaf (5 possible states)	Light green: Band across leaf; Large central area; Small central area White: Large converging bands; Small converging bands
Petiole and runner hairiness	None; Few; Many; Very many
Stipule colour	Mostly red; Some red; Red veins; No red
Calyx colour	Mostly red; Lower quarter green; Half green, half red; Mostly green; Pink lobes
Corolla; colour of petals	White; White with pink stripes; Pink
Seed colour	White; Other

The seed crop

Subterranean clover is an annual which is unique in that the ripe seed heads turn downwards and are pressed into the soil by the toughened peduncle. Although annual, it requires a cold period to induce flowering after which it flowers in longer days. However, it is not generally frost hardy. Because the seed heads are buried in the soil it is impossible to prevent reseeding once a crop has been established in a field. This has two implications for the seed grower:

1. Once subterranean clover has been established it is very difficult satisfactorily to eliminate it; dormancy and hard seededness can both occur. It is thus difficult to change seed production from one cultivar to another on the same farm.
2. Seed taken from the same field in successive years will be a mixture of generations. It is therefore impossible to operate a system of generation control as an aid to the preservation of genetic quality. In these circumstances it is necessary to rely upon the identification of off-types in the growing seed crop to ensure satisfactory cultivar purity.

Isolation

Subterranean clover is self-fertilised and isolation sufficient to prevent mixture at harvest is all that is required, i.e. a physical barrier or gap of 3 m.

Previous cropping

New cultivars should preferably be sown on land which has not previously carried subterranean clover. Otherwise an interval of at least 6 years free from plants of the species is required; this interval may be shortened if a good regime of chemical control is practical.

Difficult weeds

The seed of subterranean clover is larger than most other clovers. However, weeds listed under red clover can also be troublesome in subterranean clover.

Crop layout

For seed production a pure stand is usually best, the seed being drilled in close-spaced rows at a seed rate of 6–10 kg/ha. Companion grasses may improve the grazing, but unless carefully controlled may suppress the subterranean clover and reduce seed yield.

Crop management

To discourage the plants from burying the seed heads in the soil it is preferable to have a flat, firm surface on the field. Apart from good seed-bed preparation, grazing and cultural treatment should be designed to achieve this end. In particular, the activities of underground animals, such as moles, should be discouraged and stones should be rolled firmly into the soil. Adequate calcium, phosphorus and potassium should be supplied. The crop is sown in the autumn, and seed is normally taken from the first growth the following year. If a companion grass has been sown this should be controlled by grazing. After harvest the field can be cleaned up and prepared for a further seed harvest; some grazing will be possible, but in dry weather sheep will dig out the buried seed, and if carried to excess this can reduce the life of a stand. Care should also be taken not to move sheep too quickly from one cultivar to another as seed may be transferred.

Seed-crop inspection
The best time for inspection is when the crop is flowering. The OECD Scheme for Subterranean Clover and Similar Species (OECD 1977) requires that fields shall not be taken for certified seed if there is present more than 5 per cent of other cultivars or species with similar seeds.

Harvesting
Dense stands are easier to harvest than sparse ones. In thin stands the soil surface is exposed and this causes cracks in which much seed can be lost. In a dense stand on a well-formed surface most of the seed will be above but near to the soil surface. The seed is ready for harvest when the plants are dead and dry and the seed hard. It is not usually practical to combine direct, so the crop is mown as close to the ground as possible and then picked up by combine. After combining it is often useful to go over the field with a suction harvester to recover seed from the soil surface.

Seed after harvest

Drying, seed cleaning and seed treatment
Seed is usually dry when harvested, but may contain much debris, particularly soil or stones. Pre-cleaning with aspiration is generally required. Seed cleaning is as described for other clovers. Seed treatment is not usually required.

Other species of similar type

Those which are included in OECD (1977) are:

Medicago littoralis Rohde ex Loisel	Strand medic
M. tornata (L.) Mill	Disc medic
M. truncatula Gaertn.	Barrel medic

These species shed seed freely, but do not bury seed in the soil in quite the same way as subterranean clover; nevertheless the seed is harvested as described above, as a high proportion of it is retrieved from the soil surface. Two other species of *Medicago* have been harvested for seed in South Australia: *M. rugosa* Desr. (gama medic) and *M. scutellata* (L.) Mill (snail medic).

TROPICAL AND SUBTROPICAL SPECIES

Seed production of these species is comparatively recent and techniques have mainly been developed in Queensland, Australia. It is still a developing field where those interested in seed growing will need to consult current literature.

Centrosema pubescens Benth. Common name: centro

Cultivars of centro are available and one is included in OECD (1984). However, no international classification of cultivars has been published. The species is a vigorous climbing perennial growing 40–45 cm high.

The seed crop

Isolation
Centro is self-fertilised. It should be isolated from other crops of the same or similar species by a physical barrier or a gap of 3 m to safeguard the seed crop from mechanical mixture.

Previous cropping
Centro plants produce a high proportion (up to 60 per cent) of hard seed which can live in the soil for some time. It is advisable to allow an interval of 5 or 6 years free from centro before sowing a seed crop. A shorter interval is possible when special measures are taken to control volunteer plants or when the same cultivar is to be sown again.

Crop layout
Planting in rows is advised and inter-row cultivation in the early stages of growth helps to reduce weed competition. The rows may be up to 1 m apart; a wide spacing allows support fences to be erected when it is intended to harvest by hand. Seed rates vary between 4 and 8 kg/ha. A warm-water soak before sowing may reduce the hard-seed proportion. Inoculation with rhizobium is usually necessary.

Crop management
Centro will establish and grow well in fertile soils which are well supplied with calcium, phosphorus and potassium. However, response to additional phosphorus has been variable. Best seed yields are usually obtained from the first growth, but grazing or cutting can be used to reduce excess foliage or to adjust time of flowering so that harvest falls at a more favourable period for weather. Individual cultivars tend to differ in their responses to management practices in different areas, and seed growers will need to adjust management to suit local conditions.

Seed-crop inspection
Cultivars which have been commercialised appear to differ in flower colour characteristics, and it is likely that inspection at full flower will be best for assessing cultivar purity.

Harvesting
Centro ripens comparatively uniformly. When harvesting by hand in a supported crop two pickings will usually achieve the best yield. Seed is ready for harvest when the first pods open. In unsupported crops the pods are concealed in the foliage so that there is a lot of vegetative material to be harvested with the seed. For this reason, windrowing for up to a week before picking up with a combine is preferred to direct combining.

Seed after harvest

Drying

There will generally be green material (broken leaf and stem and immature pods) in the harvested seed and drying must be done immediately to avoid spontaneous heating. Seed should be dried to 8–10 per cent moisture if it is to be stored for any length of time, as above 10 per cent moulds will grow in the tropics. Usually seed can be dried by spreading it thinly on a tarpaulin or on a concrete floor. However, in very humid climates forced-air drying will be needed; the temperature of the drying air should not exceed 35 °C.

Seed cleaning

An air-screen cleaner will usually clean centro satisfactorily. It is essential to clean soon after drying to remove rubbish which may pick up moisture rapidly in humid climates.

Seed treatment

High hard-seed content may lead to very slow germination. Mechanical scarification of the seed before sowing can improve rate of germination; however, scarified seed is vulnerable, so scarification should be done as close to sowing as possible. Rhizobium inoculation should also be done as close to sowing as possible. Ants and beanflys will attack the seed and young seedlings after sowing so that an insecticide dust may be beneficial when applied to the dry seed. However, the seed must first be calcium coated (pelleted) to protect the rhizobia; pelleting is also necessary to protect the rhizobia in high-manganese soils.

Leucaena leucocephala (Lam.) de Wit. Common name: jumbie bean, white popinac

One cultivar is listed in OECD (1984). The jumbie bean is perennial, erect and shrub-like. It is self-fertilised and so requires isolation sufficient to prevent mixture only. It is normally sown in rows 60–120 cm apart and the seed rate is 15–40 kg/ha. The seed is usually hand harvested in its long, flat pods. A particular strain of rhizobium is required for inoculation. Hard-seed percentage may be high and scarification before sowing is desirable.

Macroptilium atropurpureum (DC) Urb. Common name: siratro

One cultivar is listed in OECD (1984). The species is a climbing perennial with indeterminate habit. Siratro is self-fertilised and requires isolation only sufficient to prevent mixture at harvest. Hard seed occurs in high proportions. Seed production is best from crops sown in rows and the seed rate is 2–3 kg/ha. Inoculation with the cow-pea strain of rhizobium is necessary. Best seed yields are obtained from irrigated crops grown in dry areas; some moisture stress at early flowering is reported to be beneficial (Humphreys 1979). The seed sheds very easily when ripe. If growth is very dense it is sometimes possible to allow the first maturing seed to shed into the foliage and to combine later, but the mass of herbage to be dealt with in this way can be very high. Alternatively, the early-ripening pods at the top of the plant can be harvested with the combine knife set high and the rest of the plant combined later. For maximum seed yield, hand harvesting is recommended from crops which have been grown on support fences; two or three pickings are needed. Vacuum harvesting at the end of the season may retrieve a worthwhile amount of fallen seed.

Stylosanthes guianensis (Aubl.) Swatz. Common name: stylo

There are four cultivars listed in OECD (1984). Yellow flowers appear to be normal, but one cultivar (Cook) is reported by Humphreys (1979) to have an orange standard with a central purple stripe. Stylo is perennial and self-fertilised, although a small, generally insignificant, amount of cross-fertilisation can occur. Isolation by a physical barrier or 3 m gap is generally sufficient. The proportion of hard seed can be high and scarification prior to sowing can assist establishment. Inoculation with rhizobium is not always necessary. Crops are usually sown in rows 45–60 cm apart and the seed rate is 2–3 kg/ha. Crops should be defoliated one month before flower initiation to give a level crop canopy at harvest. Defoliating too late reduces seed yield. The seed ripens relatively uniformly and shedding can occur quickly; harvest must not be delayed once seed falls from the pod when the plants are struck with the hand. The leaves produce a sticky exudate which makes harvesting difficult as it causes the seeds to stick together. Hand harvesting is not usually economic and crops are direct combined; however, combines work best at the driest times of the day and should not be used when humidity is high. Some cultivars repay vacuum harvesting.

Other perennial species of Stylosanthes

S. hamata (L.) Tumb.	Caribbean stylo
S. scabra Veg.	Shrubby stylo

Caribbean has one and shrubby stylo two cultivars listed in OECD (1984).

Stylosanthes humilis HBK. Common name: Townsville stylo

Although no cultivars are listed in OECD (1984), Humphreys (1979) mentions three Australian cultivars. Townsville stylo is an annual, in contrast to the other *Stylosanthes* species recorded above which are all perennial. It is adapted to a wide range of conditions in the humid tropics and warmer areas of the subtropics. According to Humphreys (1979), Townsville stylo will not produce seed if minimum night temperature falls to 9 °C or below. Photoperiod is also critical, and requirements vary with cultivar from 12 to 13 hours. Humphreys also reports that although Townsville stylo is a prolific seeder, seed yields can be drastically reduced by the disease *Colletotrichum gloesporoides*. The seed crops are normally sown in rows at 3–5 kg/ha. Inoculation is not usually necessary. The unhulled seed is hooked at the end and is therefore difficult to handle as the seeds stick together and will not flow freely over screens. Most seed is collected from the ground using a vacuum harvester, mechanical rotary brush or by hand brushing. Flowering extends over a long period, and the seed sheds rapidly so that direct combining or flail harvesting for subsequent threshing is not always effective. Material swept or sucked from the ground should be cleaned first using a wide-mesh screen to remove sticks and broken stalks, and then on a fine mesh to remove soil. In some cases it may be worth while to pass the material through a hammer-mill to break up the soil; this will also dehull the seed and make it easier to handle.

Other species

Those for which there are Australian cultivars available, but not listed in OECD (1984) are:

Desmodium intortum (Mill.) Urb.
Glycine wightii (R. Grah. ex. Wight et Arn) Verdcourt

Both are partly cross- and partly self-fertilised. Isolation is desirable. *Glycine wightii* has both diploid and tetraploid forms.

8 PULSE CROPS

The pulse crops are those legumes which are harvested mainly for their dry seeds to provide protein in the diet. Some are also harvested green as vegetables or for fodder (e.g. peas or beans). There are other legumes which are harvested as dry seeds mainly for oil, and these are included in Chapter 9 on oil-seeds; however, some of them also provide a source of protein. Yet other legumes are used as fodder crops, often as companions for grasses in pastures; these are included in Chapter 7.

Some pulse crops are perennial in nature but all are treated as annuals for seed-production purposes, although some require a cold period to induce flowering, for example autumn-sown cultivars of *Vicia faba*. The inflorescences are racemose and the seeds develop in a legume or pod. In some species flowering may take place over a long period so that seed ripening is not synchronized, making harvest difficult.

In some areas inoculation of the seed with rhizobium before sowing is desirable. This technique is well developed in many countries and is standard practice in some areas.

The inoculum is available commercially from specialist companies, and is usually carried in a finely ground peat. The best method of application is to dampen the peat with water to form a slurry which will adhere to the seed. Molasses or similar material may be added to the mix to cause it to stick to the seed. An alternative method is to sprinkle the seed with water and then to add the dry peat powder plus inoculum, but this is not as effective. Some growers add the dry powder in the seed drill at sowing time, but this is not generally as effective as the slurry treatment. Soil-application methods have also been developed.

PEAS AND FABA BEANS

Pisum sativum L. *sensu lato*. Common name: pea

Peas are grown for several purposes and cultivars differ in their agronomic characteristics according to the intended use for the crop, although some can be used for more than one purpose. The main categories are:

1. 'Vining peas' are harvested when the seeds are tender and are used either for canning or freezing.

2. 'Picking peas' are also harvested when the seeds are tender, and the fresh pods are used for shelling in the household, the peas being used as a fresh vegetable. Some cultivars in this category have tender pods and the whole pod is eaten.
3. 'Combining peas' are harvested for their dry seeds and in some areas may be harvested by hand. They have two main uses:
   (a) for use as human food, sometimes as processed peas or dry peas sold in packets;
   (b) for use as animal fodder.
4. 'Forage peas' are used for either grazing, ensiling or haymaking.

For the seed grower, the standards for trueness-to-cultivars will be higher for those categories where the produce is to be processed for food: canning, freezing or packets.

Classification of cultivars

Shape of starch grains in dry seed	Simple; Compound
Colour of cotyledons in dry seed	Green; Yellow
Plant growth habit	Bush; Tall
Number of nodes on stem up to and including first fertile node and the scales	Very few; Few; Medium Many; Very many
Pod parchment	Absent; Present
Pod apex	Pointed; Blunt
Seed colour at green shell stage	Light green; Other colours

Main distinguishing characteristics of cultivars

Foliage colour	Yellow-green; Green; Blue-green
Intensity of foliage colour	Light; Medium; Dark
Stem length after flowering	Very short; Short; Medium; Long; Very long
Leaflet size	Very small; Small; Medium; Large; Very large
Leaflet shape	Narrow elliptic; Elliptic; Broad elliptic; Narrow ovate; Ovate; Broad ovate
Stipule marbling before flowering	Absent; Present
Number of flowers per raceme	One; One or two; Two; Two or three; Three or more
Standard colour	White; Greenish; Pink; Reddish purple
Shape of base of standard	V-shaped; Straight; With two lobes
Wing colour	White; Greenish; Pink; Purple; Warm red
Pod Length	Very short; Short; Medium; Long; Very long
Cultivars with pod parchment only:	
Pod width, suture to suture	Narrow; Medium; Broad
Cultivars without pod parchment only:	
Width of cross-section of pod	Very narrow, Narrow; Medium; Broad; Very broad

Pod colour	Yellow; Green; Purple
Pod curvature	Absent or very weak; Weak; Medium; Strong; Very strong
Direction of pod curvature	Towards ventral part; Towards dorsal part
Number of ovules in pod	Few; Medium; Many
Seed colour at green shell stage	Light green; Medium green; Dark green
Weight per 1000 seeds (dry)	Very low; Low; Medium; High; Very high
Seed – wrinkled surface	Absent; Present
Seed – colour of testa	Colourless; Single coloured; Multi-coloured
Seed coat ornamentation	Marbling; Dotting; Both
Time of 50 per cent flowering	Very early; Early; Medium; Late; Very late

The seed crop

Pea cultivars may be day neutral or may respond to longer days quantitatively. All require vernalisation, but for some the cold requirement is additive.

Isolation

Peas are self-fertilised and adequate isolation is therefore provided by a physical barrier or a gap of at least 3 m from other peas to avoid mixture at harvest.

Previous cropping

The normal requirement to prevent volunteer pea plants arising from previous crops is a 2-year interval. However, peas are susceptible to several diseases and a nematode which are soil borne, and in areas where these diseases are prevalent intervals of four or five years without peas are advised to minimise the build-up of serious infection levels.

Difficult weeds

Most weed seeds are easily removed from pea seed so that normal weed-control measures should be adequate for a seed crop. However, wild oats (*Avena fatua*, *A. ludoviciana* or *A. sterilis*) may cause a seed-cleaning problem if the seed crop has suffered attack from pea-moth larvae (*Cydia nigricana*); these larvae leave holes in the peas in which wild-oat seeds become lodged. It is therefore advisable to ensure that wild oats and other grass weeds with long, thin seeds are eradicated before sowing the seed crop. In the previous autumn before a pea seed crop is to be sown in the spring, glyphosate or TCA are effective; pre-drilling in the spring tri-allate or propham can be used. After emergence of the pea seed crop alloxydium-sodium is safe to use as is dichlorprop-methyl (UK recommendations). As local conditions may vary it is advisable to seek expert advice locally.

Crop layout

The yield of pea seed is very dependent upon achieving the correct crop density. Optimum plant population depends upon the growth characteristics of the cultivar, but is generally about 75 plants/m^2 – less for large-leaved plants and higher for those with smaller leaves; there are now cultivars available which have modified leaves (the leaves are converted to tendrils and the stipules may be reduced in size (Hedley *et al.* in Jones and Davies 1983) and for these 100 plants/m^2 is desirable. To achieve these populations it is necessary to sow in rows placed close together: inter-row spaces are usually 10–12 cm, to a maximum of 20 cm. Seed rate is critical. Pea seed is relatively expensive, and unlike most other crops the cost of seed forms a high proportion of the cost of growing the crop. The seed grower should therefore calculate seed requirements carefully. Cultivars differ in 1000 seed weight, and there are differences between seed-lots of the same cultivar. To calculate seed rate, therefore, we need to know both 1000 seed weight and percentage germination. We also need to estimate a 'field factor' being 100 less the anticipated field loss which can vary from 5 to 20 per cent. Seed rate required can then be calculated from the formula:

$$\frac{\text{Population target per m}^2 \times 1000 \text{ seed weight (g)} \times 100}{\text{Percentage laboratory germination} \times \text{field factor}} = \text{Seed rate (kg/ha)}$$

(from NIAB 1982)

Pea seed crops should be sown as early as possible in the spring to achieve maximum seed yield, but at this time field losses are high. If sowing is delayed the soil will become warmer and field losses generally less so that seed rates can be reduced somewhat, although the final seed yield is likely to be less. In addition to the normal germination test there is the 'electro-conductivity test' which distinguishes seed samples which produce less vigorous seedlings. The test estimates the amounts of electrolytes which are leached from the seeds when soaked in water by measuring the conductivity of the water. Those samples which release larger quantities of electrolytes will produce weaker plants in the field if sown early when the soil is somewhat cold and wet and should therefore be avoided. The test is available from most seed-testing laboratories, and seed merchants usually use it to select the best lots for use for early sowing.

Crop management

Pea seed crops generally require management similar to that given to a food crop. Seed of vining peas is often not grown as a seed crop but is harvested from crops which have become too mature to provide good material for freezing or canning. The seed grower, however, should observe the precautions outlined in earlier chapters to preserve the genetic quality of the cultivar. Also, every effort should be made to avoid or combat pests or diseases which may cause damage to the seeds within the pod.

Seed-crop inspection

The best time to inspect crops is when flower colour can be seen. Coloured-flowered or 'tare-leaved' rogues are particularly objectionable in white-flowered cultivars used for vining or dry pea production as food crops, and some roguing may be required to remove these.

Harvesting

Small areas can be harvested by hand and placed on racks or tripods to dry before threshing; this method may be used in the earlier stages of multiplication. Larger areas are usually combined direct using lifters at the cutter bar for lodged crops. Windrowing is also possible, but is not favoured in areas where weather is unpredictable as the standing crop usually suffers less damage from rain. On stony ground a windrow may pick up stones which can cause damage to equipment. Crops can be windrowed when moisture content of the peas is around 40 per cent. At this stage the peas are firm but not hard; the lower leaves will have turned yellow. For direct combining or threshing from the windrow moisture content should be 25 per cent or less. Peas are very susceptible to handling damage when too moist or too dry and such damage (bruising or rupture of the seed coat) can ruin the germination. Therefore, combining should take place when the moisture content is as near as possible to 25 per cent and not when it has fallen below 14 per cent. To hasten the ripening process chemical desiccants can be used on the standing crop. Diquat has been shown to be safe for this purpose and does not damage germination. The desiccant is usually applied by ground sprayer and the wheelings will be badly damaged; therefore a wide-boom sprayer is preferred. The chemical is applied at the windrow stage when the seeds are at about 40 per cent moisture content, and takes a few days to work. Combining should be completed within 7–10 days of application, as otherwise inclement weather may damage the crop. To estimate moisture a representative sample of pods should be collected and the peas removed from the pods for testing. At higher moisture contents it is preferable to use an oven test as meters are not generally accurate. Care must be taken to discount the effects of showers or dew which can cause a higher reading which is only temporary.

Combines must be set carefully to avoid as much damage as possible: as slow a drum speed and as wide as concave clearance as possible. It pays also not to travel too fast and to avoid combining when the crop is wet after dew or a shower as the green material will then break up in the drum and deposit a sticky mess on the seeds; this spoils the appearance and can also encourage mould growth.

Seed after harvest

Drying

Peas harvested above 14 per cent moisture will require immediate drying to that figure for storage to the next sowing season. As pea seed is quite large it takes time for moisture to transfer from the centre of the seed to be evaporated from the seed coat; the large seed also has a rather low resistance to airflow so that air moves directly upwards and does not disperse sideways in a heap or bin. Drying must therefore be slow, and if initial moisture content is high (25 per cent or above) it is best to dry twice, allowing a short resting period between. When using ventilated bins or drying floors the seed should be turned during drying to ensure that the top layers do not become too moist which encourages mould growth. Temperature of the ventilating air should not exceed 5 °C above ambient. For continuous-flow driers the drying air should not exceed 43 °C and this should be reduced below 38 °C if moisture content of the seed is 25 per cent or above.

Seed cleaning

The main cleaning can be done on an air/screen cleaner. For stained peas an optical sorter can be used, but the output from these machines is low; more than one may therefore have to be included in the cleaning line to maintain output capacity. Cracked seed or seed which has been damaged by pea moth can be removed on a pin drum. However, when weed seeds lodge in the holes these seeds are very difficult to remove and cleaning losses can be high. Pea seed should be handled as carefully as possible as it is very prone to crack or split. Careful attention to the way in which hoppers or elevators deliver seed so as to avoid long drops and minimise impact of falling seed can prevent many of these difficulties. Equipment should not be overloaded to cause friction between the seed and moving parts. Augers are particularly objectionable.

Seed treatment

The main disease which can be seed borne is foot rot and leaf and pod spot caused by *Ascochyta pisi* which usually occurs with two other fungi: *Mycosphaerella pinodes* and *Phoma medicaginis* var. *pinodella*. The best prevention is to use clean seed and to avoid too close a rotation. If treatment is required, thiabendazole has given some protection as also has benomyl. 'Damping off' of the developing seedling caused by *Pythium* spp. is not seed borne but occurs in soil, particularly when conditions are adverse; seed treatment with captan or thiram is effective. Captan can also be obtained in mixture with thiabendazole as protection against both diseases.

Vicia faba L. Common name: field bean; broad bean; faba bean

Faba beans are grown for several different purposes, and some authorities have divided the species into subspecies or botanical varieties depending mainly upon seed size to distinguish them. However, modern cultivars no longer fit well into such groupings because seed sizes have been altered and there is not a good division into discrete classes. The main purposes for which faba beans are grown are as follows:

1. 'Broad beans' are harvested when the seeds are tender for use as a vegetable, either fresh or after freezing or canning. In some areas these may be treated as a pulse crop, i.e. harvested when mature and the dried seed eaten later.
2. 'Field beans' are harvested when mature and the dried seed is used as a stock food. Field beans are sometimes further subdivided into the 'tic beans' which have small seeds and 'horse beans' which have larger seeds.
3. Faba beans may also be harvested as whole plants for stock food, either hay or silage. Field beans are generally used for this purpose.

For seed growing the standards for trueness-to-cultivar will be higher for broad beans than for field beans, particularly when they are intended for freezing or canning.

Classification of cultivars

Melanin spot on wing petal	Absent; Present
Colour of testa of ripe seed	Beige; Green; Red; Violet; Black

Main distinguishing characteristics of cultivars

Plant height	Very short; Short; Medium; Long; Very long
Time of 50 per cent flowering (one flower/plant)	Early; Medium; Late
Standard petal anthocyanin	Absent; Present
Pod length (without beak)	Very short; Short; Medium; Long; Very long
1000 seed weight	Very small, Small; Medium; Large; Very large

The seed crop

Faba beans are either day neutral or respond qualitatively to longer days. There is also a quantitative cold requirement. Some cultivars are winter hardy and are sown in the autumn; others are sown early in the spring when there is still opportunity for the vernalisation requirement to be satisfied.

Isolation

Vicia faba is partly self- and partly cross-fertilised. The flowers are adapted for insect pollination and cross-fertilisation is mainly effected by bees. Smaller fields up to 2 ha should be isolated by 200 m when the seed to be produced is for further multiplication, or 100 m when it is not; for fields over 2 ha the corresponding distances are 100 m and 50 m. However, these distances may not be enough when seed of broad bean cultivars is being grown as higher standards are required; in this case 500 m or 300 m would be appropriate.

Previous cropping and difficult weeds

These are as described for peas. In faba beans the danger is attack by the larvae of the bean beetle (*Bruchus rupimanus*) which leaves holes in which weed seeds can lodge.

Crop layout

Spacing between rows is generally wider for autumn-sown crops than for spring-sown, with 30–35 cm as a good compromise and 20 cm for spring. However, good results have been obtained from spacings from 12 to 60 cm for autumn sowing and 12 to 50 cm in spring. Wider spacing may be beneficial in drier areas or where inter-row cultivation is practised. Seed rates vary considerably; for larger seeded cultivars such as broad beans 250 kg/ha or more may be required, while for those with small seeds such as tic beans 175–200 kg/ha will be sufficient. Lower rates are also used for free-tillering autumn-sown cultivars than for those sown in spring which tiller less. Faba beans should be sown deeper than most crops and should be planted at least 8 cm deep. Some cereal seed drills cannot penetrate to this depth and it may be necessary to adopt other methods – broadcasting the seeds and then ploughing them in or putting a seed-box on the plough are satisfactory methods often superior to drilling. Depth is required to prevent bird damage and to make possible the use of residual herbicides such as simazine.

Crop management

This does not differ from normal management, except that broad beans are harvested when the seed is mature. Great care needs to be exercised when applying insecticides close to or during flowering time to ensure that bees are not killed. Bees are the main pollinators and also 'trip' flowers to enable fertilisation to occur. For the seed grower it is usually advantageous to place hives of bees within the crop to encourage as much cross-fertilisation as possible; the resulting seeds will produce plants which are more vigorous than those derived by self-fertilisation. The normal recommendation is one honey-bee hive per hectare. In areas where the wild-bee population is high the provision of honey-bees is less necessary; however, some wild bees such as *Bombus terrestris* tend to take nectar by cutting a hole in the base of the corolla tube and this avoids pollinating the flower. Other species of *Bombus* (e.g. *B. hortorum*) visit the front of the flower and are pollinators (Broad and Poulsen in Hebblethwaite 1983).

Seed-crop inspection

The best time to inspect seed crops is during flowering. This enables rogues with different coloured flowers to be detected. However, this would be too late to correct any faults in isolation and an earlier inspection before flowering begins may also be required.

Harvesting

As with peas, small areas can be harvested by hand. The longer beanstalks can be bound into sheaves and stooked in the field to dry and to allow the seeds to mature further. Larger areas are usually combined direct. Windrowing is possible, but prolonged wet weather when the crop is in the windrow causes more damage than if the crop is left standing. Faba beans tend to ripen irregularly on the plant because flowering is not well synchronised, the plants being indeterminate. Therefore, judging the time to harvest is difficult. The maturing plant loses its leaves and then the stem and pods begin to turn black. For windrowing at least 25 per cent of pods should have changed colour; for direct combining it is advisable to wait until up to 90 per cent have changed. Moisture content will be 30 per cent or below at this time, and for direct combining should fall below 20 per cent. Premature shedding of seed is not usual except in dry, warm weather.

Desiccation is not always successful. Much of the chemical may be absorbed by the leaves, so that the stem and pods are not affected. If rain follows application the crop may be quicker to absorb moisture than if it had been untreated. However, if there is much green weed growth or if the weather is dry, desiccation with diquat will advance combining which can normally take place within 5–10 days of application. Desiccation has no effect on yield or quality. The seed is easily damaged in threshing, and careful setting of drum clearance and speed is required. Much seed can be lost by shattering at the cutter bar and reel setting is also important. In some crops it may be advisable to remove the reel altogether and to drive at greater speed so that the cut plants fall on the table. When combining lodged crops direction of travel should be against or diagonal to the direction of lodging and in some cases it may be necessary to combine in one direction only. Very ripe crops combine best when slightly damp, for example with dew to reduce shattering. Combining immature crops results in much green, damp material which can cause blockages in sieves and augers.

Seed after harvest

Drying and seed cleaning
These are as described for peas.

Seed treatment
The most serious potential seed-borne disease is leaf and pod spot, caused by *Ascochyta fabae*. The most effective control is to use only disease-free seed: infected crops should not be harvested as seed and there is also a laboratory test which can be used to screen seed-lots before they are used. Small seed quantities can be treated with thiram in warm water (37 °C); the seeds must be soaked for 12–24 hours, and even after subsequent drying they remain very tender and must be handled with extreme care. Other seed treatments are not very effective because it is difficult to achieve satisfactory penetration into the seed. Slurry treatments appear to have more promise than powders, but at present only benomyl appears to be possible.

PHASEOLUS SPP.

Phaseolus is closely related to *Vigna*, and some species previously described as *Phaseolus* are now recognised as correctly belonging to *Vigna* which is the subject of the next section. There are three species of *Phaseolus* of major importance: *Ph. vulgaris* and *Ph. coccineus* are grown both as temperate vegetables and as pulses in the tropics and subtropics; *Phaseolus lunatus* is tropical and is not tolerant of cold. One other species, *Ph. acutifolius*, is also grown as a pulse in the tropics.

Phaseolus vulgaris L. Common name: common bean, French bean, haricot bean, kidney bean, snap bean

In temperate regions *Ph. vulgaris* is grown as a vegetable and the pods are eaten when still green and tender; at this stage they may also be frozen or canned. In hotter regions they can be used as a vegetable, but more frequently are allowed to mature and the dry seed is cooked and eaten after threshing.

Classification of cultivars

Plant growth type	Dwarf (determinate and semi-determinate); Climbing (indeterminate)
Pod cross-section	Narrow elliptic; Elliptic; Broad elliptic; Cordate; Circular; Figure of eight
Pod ground colour	Yellow; Green; Violet
Pod stringiness	Absent; Present
Seed colour	Single-coloured; Multicoloured

Main distinguishing characteristics of cultivars

Leaf colour	Light green; Intermediate; Dark green
Time of flowering	Early; Medium; Late
Size of bracts	Small, Medium; Large
Colour of standard	White; Pink; Violet
Colour of wings	White; Pink; Violet
Pod length	Short; Medium; Long
Pod pigmentation	Absent; Present
Colour of pod pigmentation	Red; Violet
Degree of pod curvature	Slight; Medium; Strong
Length of pod beak	Short; Medium; Long
Seed size	Small; Medium; Large
Shape of median longitudinal section of seed	Narrow elliptic; Elliptic; Broad elliptic; Narrow ovate; Ovate; Broad ovate; Circular; Narrow kidney; Kidney; Broad kidney
Main colour of seed	White/greenish; Grey; Yellow; Buff; Brown; Red; Violet; Black
Main secondary colour of seed	As for main colour
Colour of hilar ring	Self-coloured; Not self-coloured
Resistance to *Colletotrichum*	Absent; Present
Resistance to common bean mosaic and to blackroot (virus)	Not resistant to mosaic but resistant to blackroot; Resistant to mosaic but not to blackroot; Resistant to mosaic and tolerant of blackroot.

The seed crop

Many cultivars are short day, but those selected for temperate regions are day neutral.

Isolation

Phaseolus vulgaris is normally regarded as self-pollinating, but some cross-pollination can also occur. Thus although 3 m isolation is usually enough, the more highly specialised cultivars used as vegetables for freezing or canning are isolated as for cross-fertilised plants, i.e. 100 m for crops intended for further multiplication and 50 m for other crops; these distances may be doubled when the crops are 2 ha or less.

Previous cropping
To avoid ground-keepers arising from shed seed it is generally only necessary to have an interval of 1 year between French bean crops. However, there are several fungal and other diseases or pests which attack both French beans and other legumes such as peas. To avoid building up infection it is advisable to limit these crops and a 4-year interval is usually advised.

Difficult weeds
A seed crop does not differ from a food crop in weed-control requirements and there are no major weed species which have seeds difficult to remove from French bean seeds.

Crop layout
Dwarf French bean cultivars are grown for seed in rows which may be 45–90 cm apart (George 1985), although 40 cm can be used for less vigorous cultivars. Seed rate is about 100 kg/ha, but requires adjustment in relation to seed size and germination capacity. Climbing cultivars usually have to be supported on beanpoles or string or netting stretched above the rows and supported by stakes. To accommodate this the rows have to be wider apart, up to 120 cm, and seed rates are reduced, being about half those required for the dwarf cultivars.

Crop management
This does not differ from that required for food crops. Weed control is essential to permit maximum seed yield, and to avoid difficulty at harvest. Diseases and pests also have to be controlled to ensure healthy, well-filled seed.

Seed-crop inspection
If only one inspection is possible the best time is when pod characteristics can be observed; the pods will still be green corresponding to when they are harvested for eating as a vegetable. An earlier inspection at flowering time should also be made whenever possible; plant growth habit can be seen more easily at this time. To assess disease, inspection when the seed is almost mature is best.

Harvesting
Small areas of dwarf cultivars can be harvested by hand, the plants being pulled or cut at ground level and bound into bundles for field drying. Climbing cultivars are harvested by hand-picking the pods, and if this can be done on two or three occasions a more evenly matured sample can be obtained. Larger areas of dwarf French bean cultivars can be direct combined and this is generally more satisfactory than windrowing first. Flowering may continue for some time so that pods do not mature all at the same time: usually the earlier pods produce the best seed, but some may be lost through shedding before maximum yield is reached. At maturity the pods are dry and off-white in colour, sometimes stained by saprophytic fungi. The seeds are firm. Seed moisture content of about 20–25 per cent is best, as above or below these limits the seed is very prone to handling damage. Moisture meters are not always accurate when used with French beans and assessment by the oven method is recommended. Great care must be taken when

threshing and handling the seed to avoid damage which can seriously reduce germination.

Seed after harvest

Drying
The produce of small areas which are harvested by hand can be dried by spreading the seed thinly on a well-ventilated floor, or in hotter climates on a drying floor but avoiding direct tropical sunlight. If warm air is used temperature should not exceed 38 °C. The seed must be handled carefully during drying, and if bin storage is used movement from bin to bin should be kept to a minimum.

Seed cleaning
The requirements are similar to those outlined for peas. Colour sorters are useful for beans in ensuring that the seeds are of the correct colour.

Seed treatment
This is not always effective, so the use of clean, healthy seed is particularly important. Benomyl has given some protection against *Ascochyta* and thiram or captan against *Pythium* spp.; dieldrin has been used against bean seed fly (*Delia cilicura*).

Phaseolus coccineus. L. Common name: runner bean

Classification of cultivars

Plant growth type	Dwarf; Climbing
Flower colour	White; Pink; Red; Red standard with white wings

Main distinguishing characteristics of cultivars

Pod length	Short; Medium; Long
Pod string	Absent; Present
Dry seed colour	White; Pale violet/purple with sparse black spots; Pale violet/purple with black mottle; Light tan with brown mottle; Black

The seed crop

The runner bean is used in a similar manner to the French bean, but most of the cultivars are climbing and require supports. There are some dwarf cultivars. The cultivars are long day and sensitive to temperature; seed set and pod development are best at temperatures between 20 and 25 °C. Bees or other insects are required to trip the flowers before seeds can develop, and although the runner bean can self-fertilise, out-pollination also occurs up to 40 per cent (George 1985). It is thus advisable to isolate seed crops by at least 100 m. Other seed-growing requirements are similar to those described for French beans.

Phaseolus lunatus L. Common name: lima bean; sieva bean

Classification of cultivars

Plant growth type	Dwarf; Climbing
Seed size	Small (sieva bean); Large (lima bean)
Seed colour	White; Coloured

Main distinguishing characteristics of cultivars

Leaflet shape	Round; Ovate; Ovate-lanceolate; Lanceolate; Linear-lanceolate
Colour of flower wings	White; Light pink; Deep pink to purple; Violet
Pod curvature	Straight, Slightly curved; Curved

Cultivars may also be distinguished by seed colour and seed-coat pattern which have been analysed in detail by IBPGR (1982).

The seed crop

According to Purseglove (1984), lima beans can cross-pollinate up to 18 per cent so that isolation of 100 m is desirable for a seed crop. The crop is tropical, and will not tolerate frost. Germination will not take place below 15.5 °C and some cultivars fail to set seed satisfactorily when the temperature is above 26.5°C. Requirements for the seed crop are similar to those described for French bean.

VIGNA SPP.

There are four species of *Vigna* of importance as tropical and subtropical pulses: *V. unguiculata*; *V. radiata* (= *Ph. aureus*); *V. mungo* (= *Ph. mungo*); and *V. aconitifolia*. In addition, *V. angularis* (= *Ph. angularis*) is grown in some areas.

Vigna unguiculata (L.) Walf. Common name: cow-pea

Classification of cultivars

Plant growth type	Dwarf (determinate and semi-determinate); Climbing (indeterminate)

Main distinguishing characteristics of cultivars

Terminal leaflet shape	Globose; Subglobose; Subhastate; Hastate
Plant hairiness	Glabrous; Pubescent; Hirsute
Pod attachment to peduncle	Erect; Intermediate; Pendant

Immature pod pigmentation	Absent; Present
Pod curvature	Absent; Slightly curved; Curved; Coiled
Pod length	Short; Medium; Long
Seed shape	Kidney; Ovoid; Crowder; Globose; Rhomboid
Testa texture	Smooth; Rough; Wrinkled

Additionally, IBPGR (1983) described 'eye patterns', i.e. the shape of the pigment pattern which surrounds the hilum and also 'eye colour'. It is stated that these characteristics are in use at the International Institute of Tropical Agriculture, Ibadan, Nigeria.

The seed crop

Isolation

According to Purseglove (1984) the degree of cross-pollination varies considerably between different areas, being greater in the more humid areas. Steele (in Simmonds 1984) reports up to 2 per cent outcrossing. Some isolation is therefore desirable in the early generations of seed production up to and including crops producing basic seed.

Previous cropping

An interval of 1 year free from cow-peas is usually sufficient to avoid ground-keepers in a seed crop. However, longer intervals may be desirable to avoid the build-up of disease in a field.

Difficult weeds

A seed crop does not differ from a food crop in weed-control requirements.

Crop layout

Climbing cultivars require supports, and are grown in rows 75–100 cm apart with 10–15 cm between plants. Dwarf or semi-erect cultivars are planted in rows 30–60 cm apart. Seed rates vary between 20 and 50 kg/ha.

Crop management

This does not differ from that required for food crops. Cow-peas are tropical, and grow best between 20 and 35 °C; they will tolerate 15 °C, but not below. The crop responds well to phosphorus.

Seed-crop inspection

This is most effective when pod characteristics can be observed and when it is possible also to distinguish some seed characteristics.

Harvesting

Small areas are usually harvested by hand, and picking is essential for climbing cultivars grown on supports. If the pods mature over a wide interval it will be best to pick more than once. The pods are ready for picking when they are yellowish brown and the seeds are firm. For larger areas it is best to windrow the crop to allow the pods to dry out before picking up with a combine or carting to the thresher.

Seed after harvest

Drying and seed cleaning

Small quantities of seed will normally be dried by spreading on a drying floor, which, if covered, should be well ventilated. If artificial drying is used the air temperature should not exceed 35 °C. Seed-cleaning requirements are similar to those described for peas.

Seed treatment

Treatment with captan or thiram has been used against anthracnose (*Colletotrichum lindomuthianum*), wilt (*Fusarium oxysporum*) and damping off (*Pythium* spp.). Streptomycin sulphate has controlled bacterial blight (*Xanthomonas vignicola*). However, the best way to combat seed-borne diseases is to use only healthy seed from a clean crop.

Vigna radiata (L.) Wilczech var. *radiata*. Synonym: *Phaseolus aureus*. Common name: mung bean; green gram (Fig. 8.1)

Classification of cultivars

Seed colour	Yellow; Greenish yellow; Light green; Dark green; Brown; Mixed
Plant growth pattern	Indeterminate; Determinate

Fig. 8.1 Mung bean

Main distinguishing characteristics of cultivars

Hypocotyl colour	Green; Greenish purple; Purple; Mixed
Terminal leaflet shape	Deltoid; Ovate; Acute; Ovate-lanceolate; Luneate; Lobed
Terminal leaflet length	Short; Medium; Long
Leaf pubescence	Absent; Present
Leaf colour	Light green; Green; Dark green
Petiole length	Short; Medium; Long
Length of peduncle	Short; Medium; Long
Raceme position	Mostly above canopy; Intermediate; No pods visible
Calyx colour	Green; Greenish; Purple
Corolla colour	Light yellow; Deep yellow; Greenish yellow
Colour of ventral suture	Light green; Dark green; Purple
Mature pod colour	Straw; Tan; Brown; Black
Mature pod shape	Semi-flat; Round
Attachment of mature pod to peduncle	Pendant; Intermediate; Angle of about 90°
Pod pubescence	Glabrous; Intermediate; Heavily pubescent
Pod constrictions	Absent; Present
Pod curvature	Least curved; Intermediate; Most curved
Mottling on seed	Absent; Light; Medium; Heavy
Lustre of seed	Dull; Shiny
Seed shape	Round; Oval; Drum
Hilum	Concave; Not concave

The seed crop

Isolation

According to Purseglove (1984) the mung bean is fully self-fertile and almost entirely self-pollinated. Therefore isolation sufficient to prevent mixture at harvest (3 m gap) is all that is required.

Previous cropping

One year free from other pulses is desirable. Disease is less of a problem in mung bean than in some other pulse crops.

Crop layout

Mung bean is more erect than some other pulses and there are no climbing cultivars. The crop can be planted in rows 20–45 cm apart at a seed rate of 10–12 kg/ha.

Crop management

Mung bean is grown in the tropics or subtropics at temperatures of 30–35 °C, it is drought tolerant and day neutral or short day. There are no special requirements for a seed crop different from a food crop.

Other aspects

Seed-crop inspection, harvest, drying, seed cleaning and treatment are as described for cow-pea.

Vigna mungo (L.) Hepper Synonym: *Phaseolus mungo.* Common name: black gram

Classification of cultivars

According to Purseglove (1984) cultivars are grouped on seed characteristics: those with larger black seeds and those with smaller olive-green seeds; the former mature earlier. No detailed descriptors have been published.

The seed crop

Seed-crop requirements are as described for mung bean. The pods turn black when mature.

Vigna aconitifolia (Jacq.) Marchal. Synonym: *Phaseolus aconitifolius.* Common name: mat or moth bean

Classification of cultivars

Seed colour (background)	White; Cream; Brown; Grey
Plant growth pattern	Indeterminate; Determinate

Main distinguishing characteristics of cultivars

Plant growth habit	Erect; Intermediate; Prostrate
Tendency to twine	None; Slight; Moderate; Much
Terminal leaflet base shape	Narrowly cuneate; Broadly cuneate
Lobing of terminal leaflet	None; Shallow; Intermediate; Deep
Tip shape of terminal leaflet	Round; Subacute; Obtuse
Lobe shape of terminal leaflet	Lanceolate; Broad ovate; Ovate; Rhombic
Leaf pubescence	Glabrous; Intermediate; Pubescent
Petiole pubescence	Glabrous; Intermediate; Pubescent
Days to 50 per cent flowering	Early; Medium; Late
Raceme position	Above canopy; In upper canopy; Throughout canopy
Pod attachment to peduncle	Erect; Horizontal; Pendant
Pod pubescence	Glabrous; Intermediate; Pubescent
Pod curvature	Absent; Slight; Curved (sickle shaped)
Mature pod colour	White; Cream; Brown
Seed shape	Globose; Ovoid; Elongated; Cubic to oblong; Kidney; Drum
Seed size	Small; Medium; Large
Hilum shape	Plain; Concave

The seed crop

Seed-crop requirements are as described for mung bean. The crop can be grown at somewhat wider spacing between rows, and the seed rate is generally less (about 4 kg/ha).

Vigna angularis (Willd) Ohwi and Ohashi. Synonym: *Phaseolus angularis* (Willd) White. Common name: Adzukibean

This species differs from the others which have been described because it is largely cross-fertilised. Isolation is therefore required for seed crops.

OTHER PULSES

Lens culinaris Medic. Synonym: *L. esculenta* Moench. Common name: lentil; split pea

Classification of cultivars

Seed size	Small; Medium; Large
Pod size	Small; Medium; Large
Pod shape	Flattened; Convex
Flower ground colour	White; White with blue veins; Blue; Violet; Pink

Main distinguishing characteristics of cultivars

Seedling stem pigmentation	Absent; Present
Leaf pubescence	Absent; Slight; Dense
Leaflet size	Small; Medium; Large
Tendril length	Rudimentary; Prominent
Time of flowering	Early; Medium; Late
Pod pigmentation	Absent; Present
Testa ground colour	Green; Grey; Brown; Black; Pink
Testa pattern	Absent; Dotted; Spotted; Marbled
Testa pattern colour	Absent; Olive; Grey; Brown; Black
Cotyledon colour	Yellow; Orange/red; Olive-green

The seed crop

The lentil is a native of the eastern Mediterranean countries; in tropical or subtropical areas it is cultivated at higher altitudes and in the cool, dry season. It is not tolerant of waterlogging but does relatively well during drought.

Isolation
Although largely self-fertilised, some out-pollination can occur. Early generations in a maintenance breeding programme should therefore be safeguarded, but for the late generations 5–10 m is considered to be sufficient.

Previous cropping
Lentils are usually grown in a rotation with other crops. An interval of two seasons is advisable before sowing a seed crop. In some soils inoculation with a suitable rhizobium may be necessary.

Difficult weeds
Orobanche may be a problem in some areas; seed crops should not be grown on infested land. Other legumes such as *Lathyrus* and *Vicia* species can be troublesome ground-keepers which are difficult to weed in a lentil crop since their appearance is similar in the vegetative phase; fields which have recently grown such crops should be avoided.

Crop layout
Seed crops are best grown in rows 25–35 cm apart when a seed rate of about 55 kg/ha is required. Wider spacing up to 60 cm between rows can be used in very dry areas.

Crop management
This does not differ from that required for food crops grown as pure stands. Phosphorus is necessary and applications of 40–100 kg/ha have shown good returns; however, the developing seedlings are sensitive to contact and placement is therefore important. A small starter dose of 10–25 kg/ha nitrogen has also proved beneficial. Irrigation should be used sparingly as the plants are sensitive to excess water. At flowering time *Bruchus* spp. may cause damage; they can be controlled by insecticides such as endosulphan or methialathion.

Seed-crop inspection
This is best undertaken when flower colour can be observed.

Harvesting
The plants turn yellow when ripe and the pods are brown. In most areas the plants are pulled by hand and placed in bundles in the field to dry. Larger areas may be combined.

Seed after harvest

Drying and seed cleaning
These are as described for cow-peas.

Diseases
Some diseases are seed borne or may be present in the soil when the crop is sown; therefore, seed treatment with Benlate has been found beneficial.

The main diseases and their treatments are:

Vascular wilt, *Fusarium oxysporum*	Thiram and benomyl
Collar rot, *Sclerotinium rolfsti*	Thiram and benomyl
Root rot, *Rhizoctonia* spp.	Brassicol
Stem rot, *Sclerotinia sclerotorium* or *Botrytis cineria*	Captan or thiram
Rust, *Uromyces fabae*	Captan or thiram

Cajanus cajan (L.) Millsp. Common name: pigeon pea

Classification of cultivars

Time of flowering Early; Intermediate; Late

Growth habit	Erect (compact); Semi-spreading; Spreading; Trailing
Main flower colour	Ivory; Yellow; Orange; Red; Purple
Pod Colour	Green; Purple; Mixed or Streaked

Main distinguishing characteristics of cultivars

Stem colour	Green; Sun-red; Purple
Secondary flower colour	None; Red; Purple
Streaking of second colour	None; Few; Medium; Many
Flowering pattern	Determinate; Semi-determinate; Indeterminate
Seed colour pattern	Plain; Mottled; Speckled; Mottled and speckled; Ringed
Seed main colour	White; Cream; Orange; Brown; Grey; Purple; Black
Seed secondary colour	As for main colours
Eye colour (round hilum)	As for main colours
Eye size	None; Narrow; Medium; Wide
Seed shape	Ovoid; Globular; Cubic; Elliptical

The seed crop

Pigeon peas are tolerant of drought and are less suited to the humid tropics. They are generally susceptible to frost. Most cultivars are short day.

Isolation

The pigeon pea is cross-fertilised to a considerable extent. According to Purseglove (1985) this can vary from 5 to 40 per cent, and is normally about 20 per cent. Seed crops should therefore be isolated by a reasonable distance from other pigeon pea crops. For small fields (2 ha or less) 200 m is needed, and for larger fields 100 m when the seed to be produced is for further multiplication. For seed intended for producing food crops these distances can be halved.

Previous cropping

When grown for seed the pigeon pea is best treated as an annual and should be grown as a pure stand. Fields which have grown legumes in the last 2 years should be avoided.

Difficult weeds

Other legumes occurring as ground-keepers are difficult to eradicate. The crop is generally susceptible to competition from weeds in the early stages and the use of pre-emergence herbicides may be advisable; alachlor is reported to have been successful.

Crop layout

Earlier maturing cultivars can be grown at closer spacing than those maturing later; 50 × 20 cm as opposed to 120 × 60 cm. Seed rate for the earlier cultivars is 8–10 kg/ha, and for those which are later 12–15 kg/ha.

Crop management
This is similar to that required for a food crop in pure stand. The crop will respond to phosphate (40–60 kg/ha). Seed-crop inspection is best undertaken when flower colour can be observed.

Other aspects

Harvesting, drying and seed cleaning are similar to the description for cowpeas. Most seed crops are harvested by hand pulling the plants and allowing them to dry before threshing.

Seed-borne diseases are not considered to be important and seed treatments are not advised.

Cicer arietinum L. Common name: chick-pea

Classification of cultivars

Pigmentation of stems and leaves	Absent; Present
Hairs on stems, leaves and pods	Absent; Present
Number of leaflets per leaf	Very few (3–9); Few; Many; Very many (> 13)
Flower colour	Blue; Pink; White
Number of flowers per peduncle	Few; Medium; Many
Pod size	Small; Medium; Large
Number of pods per plant	Few; Medium; Many
Black dots on seed coat	Absent; Present
Seed shape	Angular; Irregular rounded; Regular rounded
Testa texture	Smooth; Rough; Tuberculated

Main distinguishing characteristics of cultivars

Growth habit	Erect; Spreading; Prostrate
Leaflet size	Small; Medium; Large
Plant height	Short; Medium; Tall
Plant width	Narrow; Medium; Wide
Time of flowering	Early; Medium; Late
Number of seeds per pod	Few; Medium; Many

The IBPGR have also distinguished twenty-one different seed colours.

The seed crop

The chick-pea is a temperate crop, originating in the eastern Mediterranean countries. It is now mainly grown in India during the cool season. The crop is tolerant of drought but does not grow well in the humid tropics.

Isolation
Chick-peas are usually self-fertilised, but occasional out-pollination may

occur. Except for the very early stages of multiplication they may be classed as self-pollinating, requiring separation from other crops only by a clear demarcation.

Other aspects

Other seed-crop requirements are similar to those described for pigeon peas. The seed rate for chick-peas is 50–70 kg/ha for small seed and 75–120 kg/ha for larger seed. Rows are normally spaced 30 cm apart, but can be as much as 120 cm apart. The crop responds to a small starter dressing of nitrogen (15–25 kg/ha) and to phosphorus (50–70 kg/ha) and potassium (45 kg/ha). Chick-peas are susceptible to *Ascochyta* and *Fusarium* which can be seed borne, but seed treatment is not advised; it is best to choose disease-free fields for seed crops.

Psophocarpus tetragonolobus (L.) DC. Common name: winged bean, goa bean

Winged bean is generally treated as a vegetable and is sometimes grown for its tuberous root. It is adapted to the humid tropics. There are very few cultivars available, but descriptors were published by IBPGR in 1982 and it is now a subject for plant-breeding research. IBPGR lists eighteen characteristics by which different plant forms may be distinguished, including absence or presence of tubers, leaflet shape, flower colour and pod shape and colour. The winged bean is normally grown for seed on supports (trellis, strings or netting) and is harvested by hand some 28–30 days after flowering.

Lablab purpureus (L.) Sweet. Synonym: *L. niger* Medich. Common name: lablab bean, hyacinth bean

Lablab bean is used both as a pulse and as a vegetable. It is suited to dry areas. According to Purseglove (1984) there are two botanical varieties, one with longer pods and the long axis of the seed parallel to the suture, the other with shorter pods and the long axis of the seed at right angles to the suture. However, the species is generally cross-fertilised and there is confusion between these varieties and among cultivars, some of which are long day and some short. As a field crop the seed rate is 50–60 kg/ha in rows spaced widely apart. As a vegetable it is grown on supports and for seed is harvested by hand.

LUPINS

The species

There are three species of lupin which are currently in use: *Lupinus albus* L. – the white lupin is better adapted to heavier soils and is more tolerant of alkalinity than the other species; *L. augustifolius* L. – the blue lupin is intermediate; *L. luteus* L. – the yellow lupin is better adapted to light, acid soils.

A fourth species, *L. mutabilis* is being developed as a cultivated plant in some areas. In all species there are toxic alkaloids in the seed which renders it less valuable as a protein source. Modern cultivars have been bred so as to minimise or eliminate these alkaloids, but there are still, in each species, cultivars which are described as 'bitter' or 'sweet'.

Main distinguishing characteristics of cultivars within each species

Characteristic	States
Bitter principle in seed	Absent; Present
Plant height 3 weeks after seedling emergence	Short; Medium; Tall
Plant growth habit 3 weeks after seedling emergence	Erect; Semi-erect; Intermediate; Semi-prostrate; Prostrate
Leaf colour at flower bud stage	Light green; Medium green; Dark green
Plant height at beginning of flowering	Short; Medium; Tall
Plant height at green ripening	Very short; Short; Medium; Tall; Very tall
Length of terminal leaflet	Very short; Short; Medium; Long; Very long
Flower colour	White; Bluish white; Blue; Pink; Brimstone; Chrome yellow
Colour of tip of carine	Yellow; Blue-black
Pod length (green maturity)	Short; Medium; Long
Ground colour of mature seed	White; Grey
Ornamentation on mature seed	Absent; Present
Time of beginning of flowering	Early; Medium; Late

The seed crop

Isolation

Both self- and cross-fertilisation occur. *L. augustifolius* is largely self-fertilised, but the other two species cross-fertilise to a large extent; Belteky and Kovacs (1984) suggest that the proportion of cross-fertilisation is about 40 per cent. As sweet lupins will cross-fertilise with bitter lupins of the same species, it is important to provide adequate isolation. The sweet cultivars contain less than 0.05 per cent of the bitter principle, so that a relatively small contamination can cause considerable damage. For fields over 2 ha, 100 m is recommended when the seed to be produced is to be multiplied further, or otherwise 50 m. For fields of 2 ha or less the distances are 200 m and 100 m. Lupins are attractive to bees and it will normally be an advantage to site hives in or near the seed crop.

Previous cropping

Lupin seed can survive for a long time in the soil and a satisfactory interval is needed to avoid volunteer plants in a seed crop. However, there is also a need to avoid building up disease, and it is recommended that a 4-year interval free of all leguminous crops be allowed before sowing lupins. This should be sufficient time to disperse volunteer plants from shed seed.

Difficult weeds

Lupins are very susceptible to weed competition and therefore a clean crop is essential. The use of pre-emergence residual herbicides such as simazine or monolinuron is generally worth while in a seed crop.

Crop layout and management

There is some evidence that a dense crop may produce most pods on the main stem so that ripening is concentrated into a shorter period. Early sowing in spring also helps to produce more uniform ripening. This is especially important in a seed crop because when there are many lateral branches the ripening period is protracted and some immature seed will inevitably be harvested. This requires a rather higher seed rate than normal; otherwise a seed crop is managed in the same way as a crop to produce grain.

Seed-crop inspection

It is essential to observe flower colour, and therefore seed crops should be inspected when the flowers are open.

Harvesting

The pods of *L. luteus* will drop from the plant when overripe and harvest should therefore begin before the crop is fully ripe; the pods are light brown but the seed is still soft enough to mark with a thumb-nail. This is not generally a problem with *L. albus* which can be harvested when mature, the seed being hard. The moisture content at this stage will be 18–20 per cent. Seed with a moisture content of 14 per cent or less damages very easily in threshing. Crops are usually direct combined, and for seed it is important to adjust the combine carefully to avoid damage. In some circumstances desiccation may be an advantage, but as there is some risk of reducing germination it is not generally recommended for seed crops, although Belteky and Kovacs (1984) report Russian work which contradicts this view.

Seed after harvest

Drying and seed cleaning

Seed with a high moisture content is very vulnerable and moisture should be reduced below 14 per cent for safe storage. The temperature of the drying air should not exceed 40 °C. An air/screen cleaner will normally clean seed satisfactorily.

Seed treatment

Inoculation with rhizobium may be necessary in some areas; when this is so other seed treatment is not usually desirable. To combat seed-borne fungal diseases, thiram combined with benomyl has given good results, although the benomyl tends to reduce nodulation. Butacarb has also been shown to be effective.

9 OIL-SEED CROPS

These crops are grown primarily for extraction of oil from the seed, but they also provide a valuable source of protein and the residue after oil extraction is used for this purpose. Both the oil and the residue after extraction have specific analysis requirements which make it essential carefully to preserve cultivar purity in the seed. Quite small contamination with other genetic material may render the oil of inferior quality or the protein meal contaminated with toxic substances. The seed grower, therefore, has to be particularly careful to avoid mixture or contamination from undesirable pollen.

OIL-SEED RAPE AND MUSTARD

The brassicas are widespread. They form the main oil-seed crop in the temperate climates of Europe and North America, and are also grown in Asia and Africa, but not in the humid tropics.

There are two kinds of oil-seed rape, namely *Brassica napus* var. *oleifera* and *B. rapa* (which includes *B. campestris*). In both there are autumn-sown and spring-sown groups of cultivars, but generally *B. napus* provides the majority of the autumn-sown cultivars and *B. rapa* is grown more from spring sowing. Both also provide forage cultivars which are grown for seed in essentially the same manner as the oil-seed cultivars, although the requirements as to cultivar purity are somewhat less exacting.

The two kinds are alike in their seed-growing requirements except that the amount of cross-pollination is greater in *B. rapa* than in *B. napus*.

There are some areas, particularly in the Indian subcontinent, where *B. juncea* is grown for oil. In more temperate areas it is grown as a mustard and, together with *Sinapis alba*, is grown for seed in the same manner as the oil-seed rapes. *Sinapis alba* is largely cross-fertilised.

The species and their common names

Brassica napus L. var. *oleifera*	Swede rape (including hungry gap kale)
B. rapa L.	Turnip rape

Fig. 9.1 Brown mustard

B. juncea L. Czera et Coss ex Czera	Brown mustard (see Fig. 9.1)
Sinapis alba L.	White mustard

Cultivar purity is particularly important in the oil-seed rapes and mustard. Rape oil naturally contains erucic acid which has been reduced successfully in modern cultivars to levels which are not harmful. The meal, after crushing, also contains glucosinolates which are harmful to livestock and some cultivars with a reduced content are now available. Mustard cultivars are required to give a particular pungency to the condiment.

Classification of cultivars

Descriptors published by UPOV (1977) refer to *B. napus* only.

Sowing season	Spring; Autumn; Either
Erucic acid content of seed	2 per cent or less; Over 2 per cent
Time of flowering (within each sowing season)	Very early; Early; Medium; Late; Very late

Main distinguishing characteristics of cultivars

Development of lobes in leaves	Absent or very weak; Weak; Medium; Strong; Very strong
Length of main stem	Short; Medium; Long
Petal colour	Pale yellow; Yellow; Orange

The seed crop

Isolation

Brassica napus is usually self-pollinated, but trials in some areas have indicated yield increases when hives of honey-bees have been placed in the crops. *Brassica juncea* is also largely self-pollinated. On the other hand *B. rapa* and *S. alba* are cross-pollinated to a large extent. Because of the need to preserve cultivar purity, avoiding contamination between cultivars with differing oil constituents, a good isolation distance is usually recommended. The EEC *Oil and Fibre Plants Directive* 1969 (EEC 1974) requires seed crops to be isolated by 400 m if the seed produced is to be multiplied further, and by 200 m if it is not. Similar distances were specified in the OECD Herbage and Oilseed Scheme (OECD 1977). It is usually beneficial to place hives of honey-bees in a seed crop at flowering time and in Australia seven to ten hives per hectare are recommended (Weiss 1983).

Previous cropping

Brassica seeds are very long lived in the soil and great care is needed to ensure that volunteer plants do not occur in a seed crop. Such plants may be a source of mixture in the harvested seed, and may cause further damage by cross-pollinating some of the crop-plants. For the earlier generations of multiplication a gap of 5 years is recommended, and for the final generation at least 3 years. Since oil-seed rape and mustard are grown commercially to provide seed for crushing, these commercial crops must be excluded from the field for the required interval to avoid shed seed. Indeed, it is advisable to exclude all *Brassica* and closely related species, even those grown for forage or as vegetables, since in addition to problems from shed seed there is a risk of disease building up in the soil.

Difficult weeds

Charlock (*Sinapis arvensis*) and wild radish (*Raphanus raphanistrum*) are two weeds which have seeds very similar in size to the *Brassica* crops included in this section. *Brassica nigra* (black mustard) is occasionally grown as a crop species, but is also a weed in some areas. These three weeds all shed seed freely and are difficult to eliminate from a seed crop. The EEC *Oil and Fibre Plants Directive* 1969 (EEC 1974) specifies a standard of not more than one seed of *R. raphanistrum* in a 10 g sample of crop seed, and a maximum percentage by weight of 0.2 of *S. arvensis*. There are some herbicides which can be used both pre- and post-emergence, but they require very careful handling and specialist advice should be sought in the area where the crop is to be grown. It is advisable and generally necessary to select clean land for a seed crop. Another weed which has seed difficult to separate from *Brassica* seeds is cleavers (*Galium aparine*). In *B. napus* this can be controlled post-emergence with carbetamide.

Crop layout and management

For a seed crop these do not differ from that required for an oil-seed crop. Forage rapes are grown for seed in the same manner as an oil-seed rape.

Seed-crop inspection

For isolation this should precede flowering but for confirmation of cultivar and cultivar purity it may be better to return when the flower colour can be seen.

Harvesting

Ripe seed sheds easily and it is thus important to harvest at the right time. The period available for harvesting when the crop is in the right condition may be quite short and often is no more than a week.
Three methods of harvesting are possible:

1. Windrow and pick up later (by combine or for threshing).
2. Desiccate and direct combine later.
3. Direct combine.

The first of these methods can be adopted for hand harvesting and the crop can be cut and tied into bundles to dry rather than leaving a loose windrow.

A crop is ready for windrowing when the plants are beginning to turn yellow and the seed is darkening in colour. The seed is firm but not hard and can be marked with a thumb-nail. Seed moisture content should not be more than 35 per cent or less than 20 per cent at this stage. Threshing from the windrows can begin when seed moisture content has fallen below 14 per cent.

Desiccation (usually with diquat) can take place at a somewhat later stage than windrowing. However, there are some risks to germination of the harvested seed and this technique is not generally recommended for seed crops although it can sometimes be used to save a very weedy crop.

Direct combining should start when the plants have changed colour and the seeds are dark; seed moisture content will be 14 per cent or less and the seeds are hard.

Care must be exercised in harvesting to avoid excessive seed loss and special equipment is available. A vertical knife fitted to a windrower or to a combine when direct combining greatly helps in separating a heavy crop. A pick-up attachment should be used on the combine rather than lifters when combining from the windrow.

Rape-seed flows very easily and is small. Combines should be sealed to prevent seed from escaping through gaps which would not admit larger seeds.

Seed after harvest

Drying

Rape-seed is very vulnerable after harvest as the moisture content is usually too high for safe storage. The high oil content of the seed means that it can deteriorate very rapidly in store when it is not dry enough. Moisture content should be less than for cereal seed, and 9 per cent or less is required for medium-term storage (up to 6 months). For longer periods it is advisable to reduce the moisture content below 7 per cent, and for samples in long-term storage less than 5 per cent is recommended.

The small seed packs more tightly than larger seeds and so resists airflow. For on-floor drying the depth should not be more than 1 m, but less than 1 m may cause pockets of undried seed to form between the air inlets (Weiss 1983). Ventilated bins should not be over-filled. Ward *et al.* (1985) suggest a maximum air temperature 5 °C above ambient for on-floor drying, and 7 °C above ambient in ventilated bins.

Continuous-flow cereal driers can be used for rape-seed but may need attention to prevent the small seed escaping through gaps. The depth of seed

Table 9.1 Maximum safe seed temperatures when drying

Moisture content of seed (%)	Maximum safe temperature (°C)
10–17	66
19	60
21	54
23	49
25	43
27	38
29	32

Source: Ward et al. (1985).

on flat-bed driers has to be adjusted so that airflow is not restricted, but bubbling is prevented. Ward *et al.* (1985) give maximum seed temperature in continuous-flow driers as in Table 9.1; note that these are temperatures of the seed and not of the drying air.

Seed cleaning
An air/screen cleaner will normally clean seed satisfactorily.

Seed treatment
Of the diseases of rape which are seed-borne, phoma or blackleg (*Leptosphaeria maculans*) is probably the most widespread. Resistant cultivars are now available and provide the most effective way to combat this disease. Treatment with iprodine gives some protection. *Alternaria* leaf spot can be controlled by this chemical, or by organo-mercurials. The insecticide lindane is also used in conjunction with the fungicide captan.

Synthetic cultivars

Hybrid cultivars have not, up till now, been successful in oil-seed rape, but synthetic cultivars have given some promising results. The basic concept is to mix a number of selected lines in given proportions and to multiply this mixture for a restricted number of generations. Except for the limitation on the number of generations, the seed growing is similar to that described for the more conventional cultivars.

OTHER OIL SEED CROPS

Glycine max (L.) Merrill. Common name: soya bean: soybean (Fig. 9.2)

Classification of cultivars

In the USA and Canada soya-bean cultivars have been divided into twelve maturity groups designated as 00, 0 and I to X. Group 00 is adapted to the longer days of higher latitudes and Group X is the latest to mature.

Fig. 9.2 Soya bean

Cultivars adapted to the higher latitudes will flower earlier if planted in short-day conditions, so that in the tropics most cultivars mature at about the same time. Further, soya beans are sensitive to temperature and moisture conditions at flowering time, so that maturity groups in North America do not necessarily apply in other areas at similar latitudes. In IBPGR (1984) five maturity groups which correspond to the USA/Canadian groups are given as follows:

IBPGR Group 1	USA/Canadian Groups 00, 0
IBPGR Group 3	USA/Canadian Groups I, II
IBPGR Group 5	USA/Canadian Groups III, IV
IBPGR Group 7	USA/Canadian Groups V, VI, VII
IBPGR Group 9	USA/Canadian Groups VIII, IX, X

Nine groups which have not been related to the USA/Canadian groups are given in UPOV (1983): 1 very early; 2 very early to early; 3 early; 4 early to medium; 5 medium; 6 medium to late; 7 late; 8 late to very late; 9 very late.

Other characteristics used to classify cultivars

Colour of hairs on plant	Grey; Tawny
Colour of flower	White; Violet
Hilum colour	Grey; Yellow; Brown; Dark brown; Black

Main distinguishing characteristics of cultivars

Plant growth type	Determinate; Indeterminate
Plant growth habit	Erect; Semi-erect; Medium; Semi-prostrate; Prostrate

Plant height at maturity	Very short; Short; Medium; Long; Very long
Shape of lateral leaflet	Lanceolate; Lanceolate to rhomboidal; Ovoid; Elliptic
Leaflet size	Small; Medium; Large
Mature pod colour	Light brown; Brown; Dark brown
Seed size	Small; Medium; Large
Colour of seed testa	Yellow; Cream/yellow; Cream; Brown; Black

The seed crop

Isolation
Soya beans are almost entirely self-fertilised. Cross-fertilisation is generally accepted as being less than 1 per cent. The only isolation required is therefore that which will prevent mixture during harvest, that is, a physical barrier or a gap of 3 m. (Fig. 9.3).

Previous cropping
An interval of 1 year free from soya beans is usually required before taking a seed crop.

Difficult weeds
Soya beans are very sensitive to weed competition, particularly from grass weeds, in the early stages of growth, and it is therefore advisable to choose only clean land for seed crops. Some leguminous species have seeds of similar size which are difficult to separate from soya-bean seed; this problem is best

Fig. 9.3 Basic seed crop of soya bean

avoided by paying attention to the previous cropping history when choosing a field.

Crop layout and management
These are not different from those required for a crop for grain production. There is some evidence that denser stands will ripen more uniformly, but this will also depend on the conditions under which the crop is grown.

Seed-crop inspection
The best time for inspection is at maturity just before harvest when most cultivar characteristics can be seen. An additional inspection at flowering time is also advantageous.

Harvesting
Soya-bean seed is physiologically mature some 2–3 weeks before the ideal harvest time. At physiological maturity moisture content is about 50 per cent, while ideally at harvest it should be 15 per cent or less to permit storage without immediate drying. However, in some climates seed will have to be harvested at about 20–25 per cent moisture content, when immediate drying is essential. Physiological maturity is indicated by the seed turning yellow; this occurs even though the rest of the plant has not changed colour to any extent. The subsequent changes are rapid and mature seed sheds very easily in many cultivars. At this stage the leaves will have turned yellow, and will begin to drop and the pods will be dry. In humid tropical areas it can be advantageous to spray standing crops with a fungicide such as benomyl to reduce fungal infection on the maturing seed.

Small areas can be harvested by hand. Threshing should not be delayed as it is generally advisable to avoid leaving plants in the field but to get the seed into good storage conditions as soon as possible.

Larger areas are usually harvested by direct combining. Careful adjustment of the combine is needed, especially if the seed moisture content is more or less than the ideal of 14 per cent. Above this figure the seed is soft and will bruise easily; below it can be cracked by too high a cylinder speed.

Desiccation can be an advantage in some areas, particularly where early frosts or broken weather may affect the standing crop. However, there is generally some yield reduction when a defoliant is applied and this has to be balanced against expected yield losses from these other causes. In general, desiccation of seed crops should be avoided unless there are good reasons for it.

Seed after harvest

Drying and seed cleaning
Soya beans are particularly vulnerable after harvest. The seed is easily damaged by rough handling and will deteriorate very rapidly in store if conditions are not correct. The first essential is to ensure that moisture content is 14 per cent or less and to maintain viability 12 per cent is advised. The seed does not store well, and even good seed will start to deteriorate after 6 months in many areas when temperature or humidity are not controlled. Drying must be done carefully, and drying air temperature

should never exceed 40 °C; when initial moisture content is high (above 20 per cent) drying must be gradual and the air temperature reduced.

Cleaning is generally best done on an air/screen cleaner and sometimes a gravity separator may also be needed.

Storage is critical. Treatment with a fungicide such as thiram immediately after cleaning can prevent fungal infection spreading during storage, but should not be used if it is intended subsequently to inoculate the seed with rhizobium.

For storage up to 6 months seed moisture content should be maintained at 10 per cent and should not exceed 12 per cent otherwise rapid deterioration will occur. If air-conditioned storage is available, it should be maintained at 60 per cent humidity and 20 °C temperature, with the seed put into store at 10 per cent moisture. An alternative is to place the dried seed in moisture-proof containers such as heavy-duty polyethylene bags which have been heat sealed.

For storing small quantities of seed, the dried seed can be stored at 5 °C satisfactorily or it can be stored in cans.

Seed treatment

Inoculation with rhizobium is usually necessary when soya beans have not been grown for some time on a field. When seed is inoculated other seed treatment is not advisable, but treated seed will not affect rhizobium already present in the soil. Treatment with thiram, captan or carboxin are all effective in reducing seed-borne fungi.

Arachis hypogea L. Common name: groundnut, peanut

Classification of cultivars

Cultivars are usually divided into two main groups designated 'Virginia' and 'Valencia'. Gregory and Gregory (in Simmonds 1984) distinguish these as subspecies, giving the subspecific names *hypogea* and *fastigiata*. The difference between the two is in flowering habit: in the 'Virginia' group there are no floral axes on the main stem and alternating pairs of vegetative and floral axes along the lateral branches; in the 'Valencia' group there are floral axes on the main stem and continuous runs along the lateral branches. In UPOV (1985) a first division is made on the basis of a 'commercial grouping' which is given three classes: 'Valencia', 'Virginia' and 'Runner'. Other classification characteristics given by UPOV are:

Time of maturity	Early; Medium; Late
General pattern of flowering	Alternate; Sequential
Seed weight per 1000 seeds	Low; Medium; High

Main distinguishing characteristics of cultivars

Plant growth habit at flowering	Erect; Semi-erect; Prostrate
Pod constrictions	Absent or very shallow; Shallow; Medium; Deep; Very deep

Prominence of pod beak	Absent or very inconspicuous; Inconspicuous; Medium; Prominent; Very prominent
Shape of pod beak	Straight; Curved
Seed; colour of uncured mature testa	Monochrome; Variegated
Cultivars with variegated testa only, colour of uncured mature testa	White to cream; Flesh; Brown; Pink; Red; Purple; Dark purple
Dormancy period of fresh matured seed	Short; Medium; Long

The seed crop

Isolation

The groundnut is self-fertilised, although bees do visit the flowers and there may be occasional cross-pollination. For seed growing, isolation sufficient to avoid mixture at harvest (i.e. a physical barrier or gap of 3 m from other groundnut crops) is all that is required.

Previous cropping

Groundnuts usually have a relatively short dormancy period so that seed in pods left in the soil soon starts to germinate. There is thus little opportunity for the establishment of volunteer plants after the initial post-harvest germination has occurred, and this can be destroyed by cultivation. An interval of 1 year free from groundnuts is therefore usually sufficient before sowing a seed crop.

Difficult weeds

There are no particular weeds which cause difficulties in seed cleaning. However, groundnuts are very susceptible to weed competition in the early stages of growth, so clean fields are necessary for a seed crop.

Crop layout and management

Seed crops should be grown in the same manner as a crop for food.

Seed-crop inspection

The most important time for seed-crop inspection is during flowering when there is the best opportunity to spot off-types.

Harvesting

It is necessary to follow pod development below ground by lifting samples, since in many cultivars the foliage remains green until after the pods have matured. When mature, pods become dark coloured inside and the seeds assume their final colour (see distinguishing characteristics of cultivars). At this stage the seeds will have a moisture content of 30–40 per cent and the crop can be lifted. Lifting can be done by hand digging or mechanically; in either method it is essential to handle the plants carefully to avoid damage to the pods. Rough treatment which causes pods to split or bruises the seed can cause loss of germination or may allow fungi access which will destroy the seed.

After lifting, the plants have to be left in the field to dry. Where hand labour is plentiful it is best to place the crop on poles or tripods, taking care to ensure that as few pods as possible touch the ground. Alternatively the crop can be windrowed, ensuring that the plants are inverted to keep the pods off the soil. Mechanical lifters can also windrow the crop in this manner.

Removal of the pods from the plants should be done when seed moisture content is 20 per cent or slightly higher. Specialist combines are available for this operation as are small stationary or portable machines. The operation can also be done by hand. Cereal combines can be adapted to depod groundnut plants but are not usually satisfactory for seed.

Seed after harvest

Drying

The pods are stored and the seed dehulled as close to sowing time as possible. In this way the seed is much less vulnerable to in-store hazards including insect or fungus attack. Seed delivered to store will normally have a moisture content of about 20 per cent, and this must be reduced below 10 per cent for safe storage. However, over-dry pods become brittle and are very easily damaged so that care must be exercised not to dry too fast or to a moisture content much below 10 per cent.

During drying air temperature should be about 35 °C and should not exceed 38 °C. Cultivars may differ in their drying characteristics, and it is necessary to check carefully during the drying process to ensure that no damage occurs. In most tropical or subtropical areas where groundnuts are grown the seed in pods can be dried in the open by leaving sacks unclosed. Shelled seed can be stored, provided it is dry and has not suffered any damage.

Seed cleaning

Trash should be removed from pods before drying and storing; an aspirator can be used. Pods can be shelled by hand or mechanically, but the operation must always be done with care to avoid damage to the seed which may lose viability. After shelling, little cleaning is usually necessary, except to remove broken pods; care must be taken not to damage the seed coat. Size grading is not usually necessary unless precision planters are to be used.

Seed treatment

Groundnut is a legume but inoculation with rhizobium is not usually necessary; if it is to be done the seed should not be chemically treated. Organo-mercurial or thiram seed treatment has given protection against fungal diseases during germination.

Helianthus annuus L. Common name: sunflower

Classification of cultivars

Increasing use is being made of hybrid cultivars using cytoplasmic male sterility, so that in addition to open-pollinated cultivars there are now inbred lines and other components of hybrids. According to UPOV (1983) a general classification can be made on the basis of five classes:

Fineness of leaf indentation	Fine, Medium, Coarse
Time of flowering	Very early; Early; Medium; Late; Very late
Natural height of plant	Very short; Short; Medium; High; Very high
Branching	Absent; Present
Stripes on seed coat	Absent; Present

Main distinguishing characteristics of cultivars

Leaf size	Very small; Small; Medium; Large; Very large
Leaf green colour	Light; Medium; Dark
Leaf blistering	Absent or very weak; Weak; Medium; Strong; Very strong
Colour of ray flower	Ivory; Pale yellow; Yellow; Orange; Purple; Red-brown; Multicoloured
Head size	Small; Medium; Large
Shape of seed side of head	Concave; Flat; Convex; Misshapen
Main colour of seed	White; Grey; Brown; Black
Mottling on seed coat	Absent; Present

The seed crop

Isolation

Sunflowers are cross-fertilised by insects and good isolation is therefore needed. For crops where the seed to be produced is intended for further multiplication a distance of 400 m is generally recommended, and half that distance when the seed is for oil or food crops. In the production of seed of inbred or hybrid cultivars isolation is particularly important, and some seed-certification schemes require greater isolation, up to 600 m, for these crops.

Previous cropping

Volunteer plants are not usually a difficulty and an interval of 1 year free of sunflower before sowing a seed crop is generally considered to be sufficient.

Difficult weeds

There are no weeds generally which cause problems at seed cleaning. However, *Orobanche* (broomrape) can attack sunflower although it is largely confined to the USSR and eastern Europe. Another parasitic weed is *Cuscuta* (dodder). To overcome *Orobanche* a long interval (4 years) free from sunflower is required. There are herbicides which will control dodder in sunflower.

Crop layout

Except when growing hybrid cultivars, seed crops do not require any different treatment from crops grown for oil. For a hybrid cultivar the correct proportions of male and female parents must be sown separately; for sunflowers equal numbers of each parent are usually required, and it is advisable to have no more than two contiguous rows of the female. Thus alternate pairs of male and of female rows are advisable. Unlike maize, which is wind pollinated, sunflowers are pollinated by insects and we have to choose the layout which encourages them to move from the male to the female flowers.

Crop management

For open-pollinated cultivars the management of a seed crop is the same as for an oil crop. For hybrid cultivar seed production similar management is required except that it is necessary to remove the male rows from the field before harvesting the hybrid seed. Pollination is an important consideration for both open-pollinated and hybrid seed production. In most areas it will be advantageous to move hives of honey-bees into or near to the crop. It is usual to have up to four hives per hectare. For hybrid cultivars it is necessary to ensure that both male and female parents flower at the same time.

Seed-crop inspection

The best time for inspection is when the crop has passed the 50 per cent flowering stage, but before it reaches 100 per cent. For hybrids this criterion is applied to the female parent, and at inspection at least half of the male parent plants should be flowering.

Harvesting

Small areas can be harvested by hand (Fig. 9.4) and it is then sometimes an advantage to harvest selectively, cutting the heads as they ripen. The heads can be cut with little stalk and taken to a drying floor where they are laid out in a thin layer. Alternatively, hand labour may cut the stalks and the plants are then stored in the field to dry. Larger areas can be direct combined. A

Fig. 9.4 Hand-harvested sunflower heads

crop is ready to harvest when the backs of the flower heads have changed colour, usually to yellow or brown; the seeds are then hard and difficult to mark with the thumb-nail. For direct combining or threshing cut heads moisture content should be about 12 per cent. In some areas there is danger of shedding if the heads become too dry. Desiccation, with, for example, diquat, can help to accelerate harvest and if done correctly will not affect germination. Modified grain combines are suitable for sunflowers; the modifications are made to the cutter bar and table to reduce the amount of green material going through the combine. One such modification is described by Weiss (1983). Sunflower seed is very easily damaged by rough treatment and therefore all equipment should be carefully set and operated to ensure a steady flow through the machinery.

Seed after harvest

Drying

Seed harvested at about 12 per cent moisture or above must be dried immediately to bring it below 10 per cent. Seed harvested at over 20 per cent moisture cannot be left even a few hours or germination will suffer. Seed as harvested will normally contain comparatively large amounts of broken heads and stalks, usually with a high moisture content. These should be removed on a pre-cleaner, preferably with both screens and aspiration. This will greatly reduce the drying problem. Drying on a floor or in ventilated bins can usually be done with unheated air. When heat is used the air temperature should not exceed 50 °C as a general rule; when dealing with seed at higher moisture contents a lower temperature should be used.

Seed cleaning

An air/screen cleaner is usually adequate for sunflower.

Seed treatment

Various fungi are carried on the seed and can cause damage in store as well as after sowing. Therefore, dressing with captan or a similar fungicide some months before the seed is needed is good practice. There are also many insect pests which attack the seed and seedlings, so that the addition of an insecticide is usually recommended.

Hybrid cultivars

In addition to the several points noted above it is also necessary to ensure that sufficient seed of both male and female parents is available. The method of calculation was described for maize in Chapter 5.

Carthamus tinctorius L. Common name: safflower

Classification of cultivars

Rosette period from emergence	Short; Medium; Long
Extent of leaf spininess	Absent or very few; Few; Medium; Many; Very many

Number of spines on outer involucral bracts	Absent or very few; Few; Medium; Many; Very many

Main distinguishing characteristics of cultivars

Location of branches on main axis	Absent; Mostly basal; Mostly on upper two-thirds; Mostly on upper third; Entire length
Angle of branches	Semi-erect; Medium; Spreading; Drooping
Internode length	Short; Medium; Long
Leaf shape	Ovate; Oblong; Lanceolate; Linear
Serration of leaf margin	Absent; Shallow; Deep
Leaf colour	Light green; Dark green; Greyish
Leaf hairiness	Absent; Few; Medium; Many
Outer involucral bracts:	
Width	Narrow; Medium; Wide
Length	Short; Medium; Long
Attitude	Closed; Open
Location of spines	Tip only; Tip and few apical; Tip and few basal; Tip and margins; Margins
Bracts enclosing head	Incomplete; Complete
Head shape	Conical; Oval; Flattened
Diameter of primary head	Small; Medium; Large
Corolla colour	White; Pale yellow; Light yellow; Yellow; Light orange base; Yellow with base and tips of lobes orange; Red-orange; Pink; Purple
Seed colour	White; Cream; Brown; Black; Grey
Seed hull thickness	Absent; Thin; Medium; Thick; Very thick
Seed size	Small; Medium; Large
Seed shape	Oval; Conical; Crescent
Pappus	Absent; Present

The seed crop

Isolation

Although basically self-pollinated, up to 10 per cent cross-pollination can occur. The amount of cross-pollination depends partly on the cultivar being grown, and partly on the conditions under which the flowers have developed. For seed crops of the earlier generations in a multiplication sequence, isolation up to 400 m is desirable. For the final generation seed crops can be isolated by a definite barrier or a gap of 3 m, although if the cultivar to be grown is prone to outcrossing a greater distance may be needed. A seed grower should take advice from the plant breeder of the cultivar to be grown. In some areas bees may increase seed yield, but this is not always so.

Previous cropping
Safflower does not usually have a long dormancy period, so that shed seed can be induced to germinate and the seedlings destroyed. A gap of 1 year free from safflower between seed crops is advisable.

Difficult weeds
There are no weeds which cause particular problems at seed cleaning. However, safflower is very susceptible to weed competition at the rosette stage and clean fields are necessary to secure a full crop.

Crop layout
Seed crops do not differ in principle from oil crops. However, the plants are usually spiny and crops are therefore difficult to walk through which can hinder proper inspection. Therefore, it is advisable to provide some gaps or walk-ways through the crop at sowing time.

Crop management
Management for seed is no different from that required for an oil crop.

Seed-crop inspection
For the best estimate of cultivar purity seed crops should be inspected when in full flower.

Harvesting
Seed should be harvested at as low a moisture content as possible, preferably between 5 and 8 per cent. The crop does not normally lodge, and seed shedding is unusual. Therefore, in most areas where safflower can be grown, it pays to wait for moisture content to decrease. When ripe, the seed is hard and can easily be squeezed from the heads. The whole plant is brittle and the bracts will have turned dark coloured. Small areas are best harvested by hand. For larger areas direct combining is preferred; this can start when seed moisture content is 10–12 per cent, but the seed will need to be dried immediately and it is better to lose another 2–5 per cent in the field if possible. In some circumstances crops can be windrowed when seed moisture content is 20–25 per cent, and this may be advisable if the crop is very weedy. Picking up from the windrow should be delayed until seed moisture is below 8 per cent. Seed requires particular care at harvest and thin-hulled cultivars are very prone to mechanical damage. Equipment should not be overloaded and speeds, both forward speed and cylinder speed, should be adjusted to ensure as gentle threshing as is consistent with winning the crop.

Seed after harvest

Drying
Seed is best dried on the plant in the field, but if it is harvested at moisture content above 8 per cent immediate drying is essential. Bin drying is normally preferred, but continuous-flow driers with an air temperature not exceeding 40 °C can also be used. For safe storage to the next sowing season seed moisture content should be below 8 per cent. Samples are best stored under refrigeration.

Seed cleaning
An air/screen cleaner will deal with most safflower seed.

Seed treatment
Safflower is prone to several seed-borne fungi. The best treatment is to apply organo-mercurial seed dressing. Thiram may be used instead. There are also numerous insect pests which attack seeds and seedlings and it is advisable to include a compatible insecticide.

Sesamum indicum L. Common name: sesame

Classification of cultivars

There are many local cultivars of sesame and an increasing number of cultivars produced by plant breeders. However, no international descriptors have yet been published although UPOV have stated that some are in preparation. Purseglove (1984) gives the main characteristics as follows:

Season of planting	
Time of maturity	
Degree of branching	
Number of flowers per leaf axil	1; 3
Number of loculi per capsule	4; 6; 8
Colour of corolla outside	White; Pink; Purplish
Colour of capsule	Brown; Purple
Colour of seed	White; Yellow; Grey; Red; Brown; Black

The seed crop

Isolation
Although basically self-pollinated a variable amount of insect pollination can occur, and according to Weiss (1983) up to 10 per cent outcrossing has been reported with up to 50 per cent in some cultivars. Seed growers should therefore provide isolation, particularly in the early generations of a multiplication sequence.

Previous cropping
Volunteer plants can usually be controlled within one year. However, there are some diseases which are soil borne and in areas where these are prevalent a longer interval before a sesame seed crop is advisable.

Difficult weeds
Grass weeds such as *Sorghum halepense* and other *Sorghum* and *Digitaria* species have seeds similar in size and shape to those of sesame. It is not possible to control these weeds in a seed crop and therefore it is essential to choose clean land.

Crop layout and management
Similar to that required for an oil crop.

Seed-crop inspection
An inspection at flowering time or when capsules have developed in the lower axils is preferable.

Harvesting
Most local cultivars of sesame have a long ripening period and the capsules mature from the bottom of the plant upwards. The capsules are dehiscent so that the plants must be cut before seed is lost from the lower capsules. The cut plants can be placed on drying racks in the field with mats below to catch the seed as it falls. After a period of drying the remaining capsules are threshed.

Recent plant breeding has produced indehiscent cultivars which can be harvested mechanically. Where ripening is protracted, the crop should be harvested by binder and stacked before threshing either with a stationary thresher or a combine; alternatively the crop can be windrowed and picked up by combine. Direct combining can be done, but some immature seed may be harvested and the capsules are difficult to thresh. The seed is very delicate and all machinery must be carefully set and used so as to avoid damage to germination. The seed is small and flows easily; therefore combines and threshers should be sealed to prevent loss of seed through gaps in the casing.

Seed after harvest

Drying and seed cleaning
Drying is not usually necessary provided the seed is harvested at about 8 per cent moisture; it should also be free from green matter after threshing. Seed cleaning can be achieved satisfactorily on an air/screen cleaner.

Seed treatment
There are several seed-borne diseases. For *Pseudomones sesami*, Neargaard (1977) suggests streptomycin 0.1 per cent and for *Xanthomones sesami*, Abavit B.

Linum usitatissimum L. Common name: flax, linseed

Classification of cultivars

Cultivars are first divided into those intended for flax fibre and those intended for linseed oil production. The former are longer stemmed and less branched. Other classification characteristics are:

Petal colour	White; Light blue; Blue; Pink; Red-violet; Violet
Capsule: ciliation of false septa	Absent; Present

Main distinguishing characteristics of cultivars

Stem length	Very short; Short; Medium; Long; Very long
Sepal dotting	Absent or very weak; Weak; Medium; Strong; Very strong

Anther colour	Yellowish; Bluish
Style colour	White; Blue
Capsule size	Small; Medium; Large
Seed colour	Yellow; Olive-green; Brown; Variegated
Time of 10 per cent flowering	Early; Early to medium; Medium; Medium to late; Late

The seed crop

Isolation

Linum usitatissimum is self-fertilised. Separation from other crops of the same kind by a barrier or a gap of 3 m is needed to prevent mixture at harvest. Where the cultivars to be grown for seed are of different kinds (i.e. for flax and for linseed) a greater distance should be provided.

Previous cropping

A gap of 1 year free from crops of the same kind should be allowed; where a change is being made from flax to linseed or vice versa a longer interval is needed.

Difficult weeds

Seed cleaning is not usually difficult for flax or linseed crops. However, the crop is a poor competitor with weeds, so a clean field is needed. Growth-regulator herbicides should be used with great care as they can cause stem distortions which can mask cultivar characteristics.

Crop layout and management

These do not differ from that required for crops for either fibre or oil production.

Seed-crop inspection

The best time to see cultivar characteristics is when the seed crop is flowering. Flowers usually open in the early morning and fade by midday.

Harvesting

The crop can usually be direct combined. Seed is ripe when the capsules are dry and the seed inside rattles when the plant is shaken. At this stage the capsules have turned a yellowy brown and the leaves have withered, but the stems may remain green. The stems are very tough and may cause difficulties by wrapping around moving parts; they are hard to cut and knives need to be sharp. Desiccation with diquat may assist ripening in some years, but should be avoided in seed crops unless absolutely necessary. The seed flows very easily and equipment should be sealed to prevent seed escape through gaps in the casing.

Seed after harvest

Drying

Requirements are similar to those outlined for oil-seed rape. Seed is harvested at 12–16 per cent moisture content and stores safely at 9 per cent or less.

Seed cleaning
An air/screen cleaner will usually clean flax or linseed seed efficiently.

Seed treatment
There are several seed-borne fungal diseases caused by *Alternaria*, *Botrytis*, *Collectotrichum* and *Phoma*. Thiram is effective against three of these, but not against *Alternaria*. Organo-mercurials are effective against *Alternaria* and *Phoma*.

Ricinus communis L. Common name: Castor

No descriptors for cultivars have yet been published by international organisations. The seeds are poisonous and must be handled with care. Both annual and perennial forms exist, but in commercial production castor is normally treated as an annual, and shorter, dwarf forms are preferred.

The seed crop

Isolation
Castor is normally cross-pollinated by wind. Male and female flowers are produced separately on the same plant. An isolation distance of 200 m is recommended, with a longer distance (up to 400 m) when the seed to be produced is intended for further multiplication.

Previous cropping
Castor seeds have a long dormancy period, and therefore an interval free of castor plants of 2–3 years is desirable before sowing a seed crop.

Difficult weeds
Normal weed-control measures for a castor-oil crop are effective for seed crops.

Crop layout and management
The same as a crop for oil production.

Seed-crop inspection
For identifying off-types inspection during flowering is recommended.

Harvesting
Similar to that needed for an oil crop. Modern dwarf cultivars do not shed seed so easily as the older cultivars. The latter have to be harvested rather earlier, and are usually hand harvested. In mechanised production special harvesting equipment is needed as described by Weiss (1983). Capsules are removed from the plants by beating and are later hulled. The capsules should be dry when harvested, and a desiccant such as diquat can be used.

Seed after harvest

Seed cleaning

Damaged seed and broken pieces of capsule must be removed at once to avoid damage in store. An air/screen cleaner will normally achieve this.

Seed treatment

There are several fungal diseases which are seed borne and dressing with thiram or captan is usually worth while.

Hybrid cultivars

All female seed-bearing and all male pollinator lines are available and hybrid cultivars have been constructed on this basis.

10 FORAGE CRUCIFERS AND FODDER BEET

FORAGE CRUCIFERS

The species and their relationships

The rapes were discussed in Chapter 9 on oil-seed crops. This chapter includes three brassicas and fodder radish:

B. oleracea L. convar. *acephala*	Fodder kale
B. napus L. var. *napobrassica* (L.) Rchb.	Swede
B. rapa L. (including *B. campestris* L.)	Turnip
Raphanus sativus L. var. *oleiferus*	Fodder radish

The cruciferous species used in agriculture are mostly cross-fertilising, and isolation during seed growing is particularly important. The following table from NIAB (1975) shows the relationship between species as it effects isolation requirements; the different sections will not cross-fertilise with one another.

Section I. All included in this section will readily cross-fertilise with one another.

B. oleracea	Brussels sprout
	Cabbage
	Cauliflowers
	Kale
	Kohlrabi
	Sprouting broccoli
	Wild cabbage

Section II. These species can inter-cross to some extent:

B. chinensis	Chinese cabbage	
B. juncea	Brown mustard	These two inter-cross readily
B. nigra	Black mustard	
B. napus	Swede, swede rape and rape-type kales	
B. rapa/B. campestris	Turnip, turnip rape	

Section III. These will not cross-fertilise with each other or with any of the other species:

Raphanus sativus var. *oleiferus*	Fodder radish
Sinapis alba	White mustard
(*S. arvensis*	Charlock)

Because there are many forms within these different species there has been ample scope for selection of cultivars within them. Almost all parts of the plant have been utilised in these different cultivars and the seed grower has a responsibility to safeguard the particular form for which the cultivar has been selected.

Maintenance of a cultivar is usually achieved by selecting mother plants within a well-grown plot of the cultivar. Selected plants may be left to seed *in situ* after removal of all other plants, or they may be lifted and replanted in an isolation seed plot or under insect-proof protection. In the latter case, blowflies have proved to be effective pollinators and may be easier to manage than bees. It is also possible to maintain and multiply the selected mother plants by vegetative propagation.

Hybrids both within and between these species have been exploited. Within species use has been made of the self-incompatible mechanism to create hybrid cultivars. In some cases seed-production problems with the inbred lines – largely caused by very poor seed yields due to inbreeding depression – have made it necessary to use an extended crossing system such as the three-way cross in order to obtain adequate quantities of seed for commercial use; such crosses are less uniform than F1 hybrids, but this is not a difficulty with fodder crops as it might be in vegetables. The problem of 'sibs' derived from self-pollination within one of the parental lines is also of lesser importance in fodder crops.

Inter-specific hybrids have been obtained in several cases. However, none has yet been developed on a commercial scale, largely because infertility renders seed production uneconomic. The most promising appears to be *Raphanobrassica*, a cross between *R. sativus* and *B. oleracea*.

Tetraploid cultivars have been obtained, and have been commercially exploited in turnips (*B. rapa/B. campestris*).

Club-root disease, caused by *Plasmodiophora brassica*, is widespread and attacks all the brassicas. There are several races of the fungus and some cultivars are resistant to some races. Fodder radish also shows resistance. The disease is soil borne and can severely reduce seed yields; care must be taken to site seed crops on fields which are free of infection.

Brassica oleracea L. convar. *acephela*. Common name: fodder kale

Classification of cultivars

International descriptors have not been published. Kales are traditionally classified into 'Marrow stem' and 'Thousand head' types. The former has a long, thick stem which is fleshy and can be eaten by livestock; it has long internodes and axillary development is restricted to the nodes on the extension growth during bolting. 'Thousand head' has shorter, thinner stems which are tougher and less palatable to stock; the internodes are short and axillary development begins at an early growth stage so producing much leaf. However, recent plant breeding, particularly of hybrid cultivars, has produced forms which do not fit easily into this traditional classification.

Characteristics which have been found useful in Europe for describing cultivars are as follows:

Leaf length	Anthocyanin coloration of leaves
Leaf width	Waxy bloom on leaves
Serration of leaf margins	
Petiole length	Anthocyanin coloration of petiole
Petiole habit (erect or not)	
Stem shape	Amount of branching
Stem colour	Plant height

The seed crop

Kale is biennial and is usually sown in midsummer for harvest as a seed crop in the autumn of the following year. If sown too early the crop may become too dense before winter and is then prone to frost damage; if sown too late the plants may not grow large enough for vernalisation to be effective. Plants should overwinter with at least fifteen leaves or leaf scars and be subjected to minimum night temperature of 5 °C or below for about 15 weeks to encourage flowering in the following spring.

Isolation

Kale is cross-fertilised, mainly by bees. Fields to produce basic seed should be isolated by at least 400 m and those to produce certified seed by 200 m. Greater distances are advised when inbred lines are being produced for hybrid cultivars, and for the maintenance seed plots of conventional cultivars. The table on page 201 shows the crops which will cross-fertilise with kale. The seed grower should note that several of these are extensively grown in gardens where plants may be left unintentionally to flower; private households thus have to be checked as well as field crops. In some areas there are zoning schemes which stipulate the cultivars which may be grown; such schemes may be voluntary, or may be legally enforced in some countries. For example, Section 33 in Part III of the UK 1964 Plant Varieties and Seeds Act contains measures to prevent injurious cross-pollination affecting seed crops of *Allium, Beta* or *Brassica* and enables the Minister responsible for its implementation to designate areas in which seed crops can be protected.

Previous cropping

Cruciferous seed can live a long time in the soil. Charlock (*S. arvensis*), for instance, is notorious for appearing in fields which have been free of the weed when cultivations are deeper than usual. Therefore at least 5 years should be allowed between cruciferous seed crops, and in the 2 years before a seed crop all crucifers should be excluded from the field. Care should be taken to eliminate cruciferous plants in the preceding crops by using appropriate herbicides.

Difficult weeds

Charlock is an obvious difficulty. The field should be carefully prepared and it is advisable to avoid deep cultivation in preparing a seed bed for the seed crop. Other crucifers will also be difficult to deal with if harvested with the kale seed, as will docks (*Rumex* sp.), fat hen (*Chenopodium album*), cleavers (*Galium aparine*), cranesbill (*Geranium dissectum*) and white campion (*Melandrium album*).

Crop layout

Seed crops are usually sown in rows 35–70 cm apart. Narrower spacing between rows is possible, but this renders roguing and the identification of off-types difficult if not impossible. Seed rate should not be too high and is usually between 1.5 and 3 kg/ha.

Crop management

In the early stages of a seed crop, management, including fertiliser application, does not differ from management of a crop for fodder. The crop is left untouched over winter. Spring application of nitrogen is not advisable unless the crop is very backward. At flowering time pollen beetle (*Meligtnes* sp.) may attack the flower buds; it invades from the hedgerows, and spraying with an insecticide is advised if the number of beetles reaches twenty per plant; before this it is often worth while to spray the headlands only to prevent build-up. Seed weevil (*Ceuthorhynchus assimilis*) also enters the crop from hedgerows and severe damage can be caused by a weevil population of one or two per plant. Control can be achieved with the same insecticide as is used for pollen beetle, and application is made at the yellow bud stage. As bees may be active in the crop at this time it is important that bee-keepers are alerted when insecticides are used so that the bees can be confined to the hive. To ensure adequate pollination it is recommended that two to three hives per hectare are sited in or near the seed crop. Low temperature at this time may reduce seed set. Fungal diseases can cause damage in a seed crop and crop hygiene is necessary to avoid build-up of infection as much as possible – suitable rotations and destruction of crop and weed debris in preceding crops are important considerations. Of the diseases which are prevalent, only downy mildew (*Peronospora parasitica*), and powdery mildew (*Erysiphe cruciferarum*) can be controlled effectively by fungicides.

Seed-crop inspection

Although vegetative characteristics are useful for checking varietal purity, it is often difficult to distinguish individual plants in a well-grown, relatively dense seed crop. Therefore crops are usually inspected when in flower, off-types being identified by flower colour and size. This also provides opportunity to check isolation as flowering brassicas in surrounding fields and gardens can be seen easily.

Harvesting

Seed ripening will take place over an extended period and it is necessary to time harvest so as to secure the highest yield of good seed. Normally the earlier flowers produce the best seed, but some shedding from these earlier-ripening pods is inevitable. Samples of pods should be collected from the middle of the stems from plants in different parts of the field. Cutting can begin when seed in these pods is light brown in colour and firm when pressed between finger and thumb. Crops are normally windrowed and allowed to dry for about 2 weeks before picking up with a combine. Windrows should not be turned unless absolutely necessary as shedding losses can be severe. Direct combining is not usually possible as there is too much green material for the combine to deal with. In some crops, however, it may be possible to set the cutter bar high so as to cut the pod-bearing branches, but this is only possible in very dense crops of relatively short cultivars. Desiccation cannot normally be used in seed crops because of the difficulty in applying the chemical in a tall, dense crop without causing undue damage and consequent seed loss.

Seed after harvest

Drying

For storage in sacks moisture content should be below 10 per cent and in bulk below 8 per cent. Harvested seed is usually above these limits and requires immediate drying to avoid damage to germination. Air temperature should not exceed 38 °C when the moisture content is below 18 per cent, or 27 °C when drying seed above this figure. When using on-floor or bin driers the seed must not be too deep as the small seed resists airflow; seed depth should generally be about half that used for cereals, provided there is enough depth to allow air distribution from lateral ducts when using on-floor driers. Seed yields are of the order of 1 tonne/ha, from which the drying capacity needed can be calculated.

Seed cleaning

An efficient air/screen cleaner should overcome most problems. A gravity separator is required in some cases and a spiral separator is also useful. The main cleaning problem is to remove light or shrivelled seed. Size grading may be required for seed when intended for use in precision drills.

Seed treatment

This is possible against dark leaf spot (*Alternaria brassicicola*) and canker (*Phoma lingam*), but in general it is preferable to aim for disease-free seed by using clean seed to sow the seed crop and maintaining suitable isolation. For *Alternaria* organo-mercurials used as a soak or iprodione have been used; for *Phoma*, hot-water treatment (50 °C for 20–30 minutes) or treatment with benomyl.

Hybrid cultivars

Several cultivars from Cambridge Plant Breeding Institute with very good agronomic performance have been produced as triple-cross hybrids. The system is illustrated in Fig. 10.1. If all seed is produced in the normal way the production of certified seed will take 11 years, but it is possible to shorten this period by vernalising the seed in the earlier stages so that the seed crop can be sown and harvested within 1 year; alternatively it is possible to multiply some of the stages on the other side of the equator so obtaining two crops in 3 years. In the production of hybrids, seed is normally harvested from both parents and mixed. To achieve this two methods are possible: the seed used to sow the seed crop can be mixed in equal proportions which automatically gives a mixture at harvest; or the two parents can be sown in separate rows so that the mixture occurs during and after harvest. The first method is the simplest, but is at a disadvantage if there is a difference in size or vigour between the two parents so that one may be unduly suppressed when growing in the mixed population. The alternative overcomes this problem, but is more complicated for the seed grower who has to ensure that

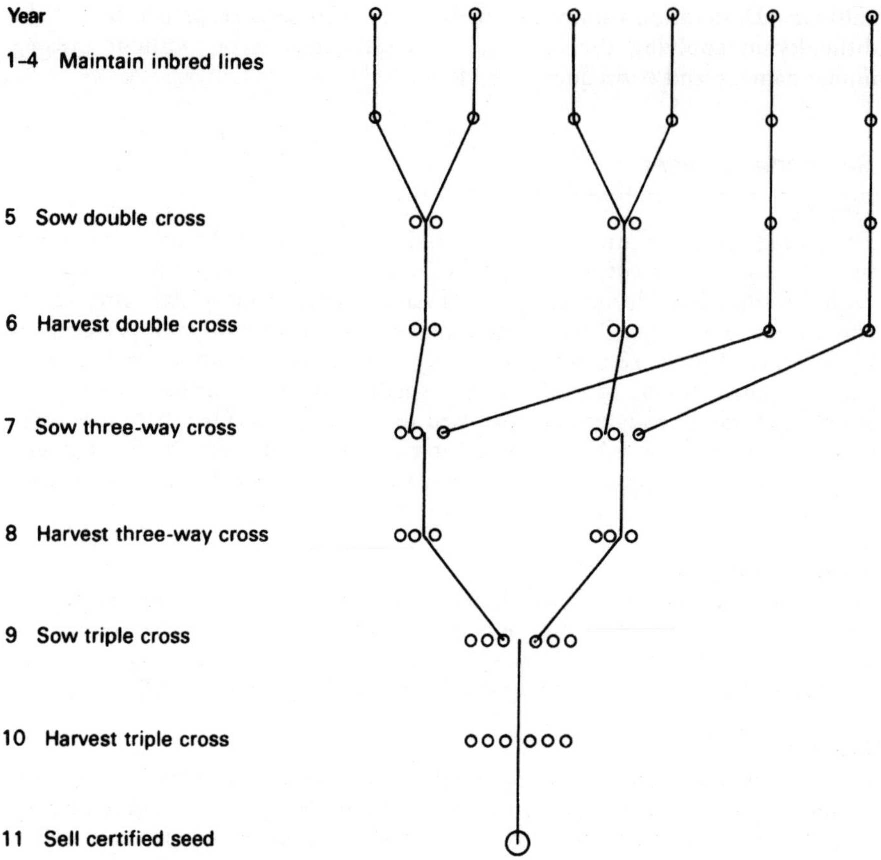

Fig. 10.1 Production of triple-cross hybrid kale

roughly the same number of plants are established in each row; experiments have shown that alternate rows of the two parents are necessary to achieve good seed set. The alternate row system has advantages for the seed-crop inspector, who is better able to distinguish the characteristics of the two parental lines when they are sown separately.

Brassica oleracea L. var. *capitata* L. Common name: cabbage

The seed crop

A small amount of cabbage is grown as a fodder crop. In the past, fodder cultivars have been used for this purpose, but recently cultivars intended for use as vegetables have been grown as fodder crops. Seed production is similar to that described for kale except that the plants in the seed crop have to be wider apart to allow the cabbage to develop. If the cabbage heart is very tight it will be necessary to slash the top of the heart so as to release the growing point for flowering. A detailed description of cabbage-seed production is given by George (1985).

Production of small quantities of kale or cabbage seed

To produce small quantities of seed for further multiplication in a conventional manner, mother plants are selected from a well-grown crop in the vegetative stage. These plants are then moved to an isolation plot during the winter and allowed to seed in the following year. Alternatively, cuttings can be taken from the mother plants and multiplied for transplanting into an isolated seed plot later in the following year; vegetative propagation may extend the period from plant selection to seed harvest unless the propagated cuttings can be vernalised.

Brassica napus L. var. *napobrassica* (L.) Rchb. Common name: swede

Classification of cultivars

Swedes were traditionally classified on the basis of root shape and skin colour. Root shape was either globe or tending towards elongated while skin colour varied from green to deep purple. In UPOV (1984) it is suggested that cultivars should be grouped as follows:

Leaf lobes	Absent; Present
Intensity of root skin anthocyanin colour above soil level	Very weak; Weak; Medium; Strong; Very strong
Colour of flesh of root	White; Yellow

Main distinguishing characteristics of cultivars

Leaf attitude	Semi-erect, Intermediate; Semi-drooping
Leaf colour	Light green; Medium green; Dark green
Leaf glaucosity	Absent or very weak; Weak; Medium; Strong; Very strong
Number of major leaf lobes	Few; Medium; Many
Size of terminal leaf lobes	Small; Medium; Large
Total length of longest green leaf, including petiole	Short; Medium; Long
Width of leaf at widest point	Narrow; Medium; Broad
Number of minor lobes on petiole	Few; Medium; Many
Thickness of petiole	Thin; Medium; Thick
Anthocyanin coloration of root skin above soil level	Absent; Present
Colour of root skin below soil level	White; Yellow; Reddish
Root shape	Transverse elliptic; Circular; Broad elliptic; Obovate; Oblong
Root length	Short; Medium; Long
Root width	Narrow; Medium; Broad

Length of neck	Short; Medium; Long
Colour of neck surface between leaf scars	Uniform red or purple; Green or purple mottled with green
Intensity of yellow colour of flesh of root	Weak; Medium; Strong
Dry-matter content of root	Low; Medium; High

The seed crop

Isolation, previous cropping and difficult weeds

These are as described for kale, although there is a higher degree of self-fertilisation in swedes than in kale.

Crop layout

Seed crops of swedes are sown in rows 35–50 cm apart and the plants are spaced in the row about 5 cm apart. Narrower spaces between the rows can be used, but this would not provide suitable plants for assessing cultivar purity. The wider row spacing allows for some inter-row cultivation where necessary. Seed rate is 2–4 kg/ha. The aim is to have a higher population than would be best for fodder production, giving a smaller root but one still sufficiently large for cultivar purity to be assessed on the basis of root characteristics.

Crop management

Early crop management is similar to that needed for a root crop, and normal fertiliser and herbicide applications should be made. The crop should be sown at a time when the plants will establish well before winter so as to complete vernalisation, but will not become too large as large plants are more susceptible to frost damage. Spring applications of nitrogen should be avoided unless the crop is very backward as excess nitrogen can delay flowering and cause excessive vegetative growth. However, late applications of nitrogen after flower bud formation may increase seed yield. Measures to combat pollen beetle and seed weevil were described for kale and apply equally for swede. Additionally, swede seed crops may be attacked by bladder pod midge (*Dasineua brassicae*), but the measures taken against seed weevil will normally also deal with this pest. Fungus diseases are as described for kale.

Seed-crop inspection

Two inspections are generally desirable: one to check the vegetative characteristics, particularly those of the roots; the second should be at flowering when isolation can also be checked.

Harvesting, drying, seed cleaning and seed treatment

These are as described for kale.

Hybrid cultivars

These have been produced experimentally but have not yet been exploited commercially. Seed production of a hybrid is relatively expensive, and although experimental hybrids have shown marked yield increases over existing cultivars, almost equally good results have been shown from conventionally bred line selections. It seems likely, however, that if hybrid cultivars are developed they will be double-cross rather than the triple-cross now used for kale.

Brassica rapa L. (including *B. campestris* L.). Common name: turnip

There is a greater variation in plant form among the cultivated turnips than in swedes. Turnips are generally quicker growing and are probably more frequently used as vegetables; there is a greater difference between the turnip vegetable and the turnip fodder plant than occurs in swedes. Apart from this greater variability, however, seed production of fodder turnip is very similar to swede, except in the points noted below; for seed production of vegetable turnip, see George (1985).

Classification of cultivars

In UPOV (1976), it is suggested that turnip cultivars may be grouped by the following characteristics:

Ploidy	Diploid; Tetraploid
Colour of flesh of root	White; Light yellow; Dark yellow
Shape of root	Transverse elliptic; Circular; Broad elliptic; Broad ovate; Oblong; Obtriangular; Ox-horn; Flask
Lobes of leaf	Absent; Present
Anthocyanin coloration in skin of top of root	Absent; Present
Colour of skin of root below ground	White; Yellow; Black

Main distinguishing characteristics of cultivars

Density of hairs on margin of first leaf	Weak; Medium; Strong
Leaf attitude	Erect; Semi-erect; Horizontal
Leaf length	Short; Medium; Long
Development of leaf lobes	Very weak; Weak; Medium; Strong; Very strong
Intensity of green colour of foliage	Weak; Medium; Strong
Position of root in soil	Very shallow; Shallow; Medium; Deep; Very deep
Intensity of anthocyanin coloration in skin at top of root	Very weak; Weak; Medium; Strong; Very strong
Green colour in skin at top of root	Absent; Present
Dry-matter content of root	Low; Medium; High
Speed of root formation	Slow; Medium; Fast
Time of flowering	Very early; Early; Medium; Late; Very late

Intensity of yellow colour in petiole	Weak; Medium; Strong
Resistance to club-root	Not resistant; Resistant (indicate race)

The seed crop

Turnips are almost entirely cross-fertilised. Bees are effective pollinators and it is worth while to introduce two or three hives per hectare in a seed crop at flowering time. The seed of turnip is somewhat smaller than that of swede, so that rather lower seed rates are required. Cultivars vary greatly in their vernalisation requirement, and some can be sown in early spring to flower the same year; the seed grower should seek advice from the breeder on this point. Some work has been done on sowing vernalised seed: the seed is first germinated and then placed in a refrigerator at 5 °C for some weeks (the exact length of time will depend upon the cultivar). The germinated seed has to be sown immediately on removal from the refrigerator, and seedling establishment may be impaired. The method provides the possibility to establish a seed crop after a poor autumn, but is not generally preferable to autumn sowing.

Production of small quantities of swede or turnip seed

By growing mature roots it is possible to select healthy mother roots for seed production. The roots are lifted from the field and stored over winter for planting in an isolation plot early in the following year.

Raphanus sativus L. var. *oleiferus*. Common name: fodder radish

Although some cultivars are biennial, fodder radish is not frost hardy and so seed crops are normally sown in early spring. Some cultivars are quick to flower and may be sown later for seeding in the same year. The very early flowering cultivars have little or no root development, whereas those which flower later produce swollen roots. Flower colour may be white, pink, purple or yellow.

The seed crop

Fodder radish is cross-fertilised and requires isolation. It will not cross with other brassicas (see Table on page 201). However, it will cross with wild radish (*R. raphanistrum*) and seeds of this weed are impossible to remove from seed of fodder radish. Therefore the field where the seed crop is to be grown and the surrounding area must be free from wild radish. Other difficult weeds are listed under kale. The seed crop is sown in rows 45 cm apart at a seed rate of about 6 kg/ha. Nitrogen at the rate of about 100 kg/ha will usually be required and is best applied at sowing time. Crop inspection at flowering time to check flower colour and isolation is considered to be most suitable. Seed is slow to ripen and is very vulnerable to birds (crops should generally be at least 4 ha to minimise bird damage). The pods turn reddish brown, and when the seed also starts to turn brown the crop can be windrowed. Drying in the windrow can take three weeks before picking up with a combine. Seed should be dried immediately to 10–12 per cent moisture.

FODDER BEET AND MANGELS

Beta vulgaris L. Common name: fodder beet; mangel

Different crops within the species

Fodder crops derived from *B. vulgaris* include the fodder beets, which have roots with a generally high dry-matter content, and mangels which have a relatively lower dry-matter content. The mangels traditionally have rather better-shaped roots which are set further out of the ground; the fodder beets have longer, narrower roots which are deeper in the ground and more difficult to lift. Recent plant breeding in the species has concentrated on the fodder beets, and cultivars are now available which have roots with high dry-matter content and of good shape which are not too difficult to lift.

Beta vulgaris also includes sugar-beet and the vegetable beetroots. Much of the plant breeding in fodder beet has come about as a development of work in sugar-beet. Thus nearly all fodder-beet cultivars are now monogerm – that is, the seed produces a very high proportion (90 per cent) of single seedlings, as opposed to the multigerm seed produced previously which gave up to three or four seedlings from each 'seed'. The 'seed' of beets is in fact a cluster of seeds which are fused together; in the genetic monogerm only one viable seed per cluster is produced.

Research on sugar-beet has shown the virtue of polyploidy in increasing yield and improving other qualities, and the same breeding systems have been applied to fodder beet. In OECD (1984) the following possible combinations are given, the first being the seed-bearing component and the second the pollen-shedding component:

(2N)	= diploid without male sterility
(2N × 2N)	= male sterile diploid with a male diploid
(2N × 4N)	= triploid with a male tetraploid
(4N × 2N)	= triploid with a male diploid
(4N)	= tetraploid without male sterility
(2N + 4N)	= polyploid without male sterility
(2N × (2N + 4N))	= polyploid with a male sterile female and a male polyploid

The seed grower thus has to know the kind of cultivar which is to be grown. For the diploid, tetraploid and polyploid cultivars without male sterility the whole crop can be taken for seed, the seed crop being sown with a mixture. When male sterility is used it may be necessary to sow the two parents separately in an appropriate ratio of female : male rows and to harvest the seed from each parent separately. Proportions of 3 : 1 or 4 : 1 have been used.

Earlier mangel cultivars were classified by root shape and root skin colour, but these characteristics are not important in modern fodder-beet cultivars.

The seed crop

Isolation

Beet is cross-fertilised. Some pollination is by insects and some by wind, but it is generally considered that wind-borne pollen is the main source likely to cause contamination. Dark (1971) reports experiments with sugar beet and recommends that breeders' seed crops should be produced in protected housing; that basic seed crops should be isolated by 1000 m and a 4.6 m (5 yd) strip on the outside of the crop discarded; and that commercial seed crops should be isolated by 400 m. He further suggests that all seed crops should be as nearly square as possible so as to present the shortest possible proportion of outside edge. The generally accepted isolation distances in Europe are:

Isolation required from:	*Basic seed (m)*	*Certified seed (m)*
Other cultivars of the same subspecies	500	300
Other subspecies of *Beta vulgaris*	1000	600

The shorter distances thus apply within fodder beet and mangels, and the longer between fodder beet and mangel and other crops such as sugar-beet and vegetable beet. Subspecies of *B. vulgaris* which will all cross-pollinate freely include: fodder beet and mangels, sugar-beet, red beet, wild (sea) beet, spinach beet, sea-kale beet or Swiss chard. In some areas there are zoning schemes designed to avoid seed crops of different subspecies from being sited too close to one another. These schemes may be voluntary or compulsory (see 'Isolation' under kale for more details).

Previous cropping

Beet seed can survive in the soil for long periods. Beet-seed crops should not be taken from the same field more often than 1 year in 6 (i.e. a 5-year interval free of beet-seed crops). Additionally, during the two years before a seed crop all beet crops should be excluded.

Difficult weeds

'Weed beet' must be avoided at all costs. These are annual forms of beet and may be wild beet or crosses between wild beet and fodder beet, or they may arise from bolters in a normal beet crop. Characteristically, weed beet sheds seed in an arable crop and can survive and multiply for many years unless dealt with. The seed grower must ensure that the field to be used for a seed crop is free of weed beet, and must also ensure that there is no possibility of contamination of the seed crop by pollen from weed beet, wild beet or bolters in adjacent beet crops. Weed beet seed cannot be eliminated from fodder-beet seed; in the UK, sugar beet seed producers make special tests for the presence of weed beet seed in seed-lots produced abroad, especially from areas where wild beet is prevalent. Other weeds which can cause difficulty during seed cleaning are:

Atriplex patula	Common orache
Chenopodium album	Fat hen
Galium aparine	Cleavers
Polygonum sp.	Bindweed
Stellaria media	Common chickweed

Crop layout and management

There are three ways in which seed crops can be grown.

1. Mature roots may be selected in the field and either allowed to seed *in situ*, other plants in the plot being destroyed, or the selected roots may be moved for seeding in isolation.
2. Seed is sown in a seed-bed to produce young plants or stecklings which are transplanted in the autumn or after overwintering.
3. Seed is sown and the plants are allowed to mature and produce seed *in situ*.

Of these three methods, the first was used for mangels when roots were selected on the basis of shape and colour. It is now rarely used with modern fodder beets and mangels with a high dry-matter content.

The second method also is now little used except during plant breeding or where labour is plentiful.

The *in situ* method is now normally used for the production of commercial quantities of fodder-beet seed. The seed can be sown either on bare ground in midsummer, or earlier under a suitable (generally cereal) cover crop. In the latter case, the cover crop must not be too dense or the seedling beet plants will be suppressed. A plant population of about 300 000 plants per hectare should be the aim: on bare ground this can be achieved with rows 19 cm apart and the plants spaced 25 cm apart. With undersown crops the rows can be closer together and the plants spaced wider apart in the rows.

Weeds can generally be controlled by chemicals, although those that can be used when undersowing will be restricted by the cover crop, and some inter-row cultivation when that crop is removed may be required.

Fodder beet requires vernalisation before it will flower although annual forms, or bolters, also occur. These should be removed from a seed crop if they become apparent. The exact length of cold period required is partly dependent upon the cultivar, and it must be followed by longer days to induce flowering.

Diseases and pests similar to those encountered in a fodder crop are also found in a seed crop, and control measures should be taken as necessary. Of particular importance is virus yellows, which can spread from a seed crop to surrounding fodder crops or to sugar-beet. The aphid vectors should therefore be controlled by suitable insecticides.

Another problem sometimes encountered is that seed may be at least partially vernalised on the plant before harvest. Such seed will be more likely to produce bolters when sown to grow a fodder crop in the following year. Therefore seed crops should not be sited in situations where cold conditions may bring accumulated low temperature prior to seed harvests. The amount of low temperature required to bring about this condition depends largely upon the cultivar being grown.

A seed crop benefits from a late application of nitrogen, generally at the rate of 200–250 kg/ha. Excess nitrogen or top-dressing too early may promote excessive growth of side-shoots which produce later maturing and smaller seed, so complicating harvest without adding significantly to the yield of good-quality seed; lodging may also occur.

Seed-crop inspection

It is difficult if not impossible to recognise cultivars in the field in a seed crop grown by the *in situ* method, because mature roots are not grown. Crop inspection has therefore tended to concentrate on disease assessment – particularly virus yellows – and checking isolation. Virus yellows assessment is made in the autumn either in the steckling bed or in the sown crop, and is intended mainly as a safeguard to eliminate badly infected crops which might carry the disease forward to the following year; thus it is primarily concerned to protect fodder or sugar-beet crops being grown in the district. Inspections at this time can also detect any contamination in the seed crop caused by red beet because anthocyanin pigmentation will be apparent. To check isolation the seed crop is visited at flowering time, particular attention being paid to waste areas in districts where wild beet might be expected.

Harvesting

Beet seed sheds easily when ripe, but because of the extended flowering period some shedding of the first ripened seed will occur before harvest. The plants change colour and when yellow with some seed lost will be ready for cutting. At this stage the plants are cut. Small areas can be cut by hand, the seed stalks tied in bundles and stooked. Larger areas are windrowed, leaving as long a stubble as possible to keep the windrow off the ground. Once cut, the crop should be handled as little as possible before threshing to avoid seed loss. Desiccants have been used to hasten drying without ill effects on germination of the harvested seed. In some instances they have been applied successfully to the windrow. Threshing by combine does not usually present any difficulty.

Seed after harvest

Drying

Seed should be reduced to 10 per cent moisture for short-term storage and to 8 per cent or below for the longer term. Drying temperature of the air should not exceed 38 °C. Batch or continuous-flow driers are the most satisfactory.

Seed cleaning

Beet seed requires to be processed carefully and is normally done by a specialist. Multigerm seed is usually rubbed and graded to achieve single seededness in about 60 per cent of the seeds. Genetic monogerm seed is carefully size graded to produce uniformly sized seed for pelleting.

Seed treatment

Most seed is treated with a fungicide and sometimes with an insecticide, although the latter is now usually replaced by distributing granules with the seed at or just before sowing time. After treatment the seed is pelleted; this operation must be performed by specialists with suitable equipment. Pelleting produces seed of uniform size which can be drilled by precision seeders to give the desired plant spacing in the field.

APPENDIX

SEED YIELDS

The yield obtained from a seed crop is influenced by many factors. Cultivars differ in their seed-yielding ability and there are many natural influences which vary widely from location to location, while the skill or otherwise of the seed grower in managing the crop will have a major effect on the end result. Total yield of seed is obviously important, but this will be subject to losses during drying and cleaning which may be anything from 10–30 per cent depending upon the efficiency of the grower in presenting the crop. Of equal importance is the viability of the seed which may be quite low in some tropical fodder crops; pure live seed yields of 25 per cent or less of the total seed yield have been quoted. Normally, however, seed yields of other crops should germinate above 80 per cent and for most cereals 98 or 99 per cent is expected.

Yields of cereals, pulses, oil-seeds and other crops where the seed is used as food or for other purposes are published annually in the FAO Production Year Book (FAO Statistics Series, Rome). These estimates of national yields are of average crops and are probably below what a good seed grower would achieve.

No such data are available for seed yields of crops grown for their leaves or other plant parts, mainly fodder crops.

Of the temperate grasses used mainly in agriculture the ryegrasses give the highest seed yields, normally of the order of 1000 kg/ha. The smaller-seeded timothy yields less than half this quantity (about 400 kg/ha) while cocksfoot, meadow and tall fescue and brome are intermediate.

The highest-yielding amenity grasses are the *Poas* and red fescue with a seed yield of about 600 kg/ha. Agrostis yields about 250 kg/ha and others are intermediate.

The prairie grasses give seed yields of about 400 kg/ha. The tropical grasses are very variable and yields of about 100 kg/ha per harvest are common, but with only about 25 per cent germination. Higher yields are normally obtained from Rhodes grass and Sudan grass.

Seed yields from forage legumes are also very variable. The temperate clovers normally yield about 350 kg/ha although yields of 750–1000 have been recorded. Lucerne will yield 500 kg/ha and the vetches up to 2000.

The legumes adapted to hot summers generally yield 300–350 kg/ha although Sericea lespedeza is higher yielding at about 750 kg/ha. Subterranean clovers and the medics yield up to 500 kg/ha.

The tropical and subtropical legumes are particularly variable. Potential yields up to 1000 kg/ha may be obtained, but at least half of the potential is normally shed seed and it is generally not possible to recover yields of more than about 600 kg/ha even with suction harvesters; yields of the order of 200–300 kg/ha are common.

Of the fodder brassicas, rape and swedes give the highest seed yields at about 1500 kg/ha; kale, turnip, radish and cabbage normally yield about half this amount. Fodder beet will yield 1000 kg/ha.

BIBLIOGRAPHY AND REFERENCES

Albeke, D. W., D. O. Chilcote and **H. W. Youngberg** (1983) Chemical dwarfing effects on seed yields of tall fescue (*Festuca arundinacea*) cv. Fawn, fine fescue (*Festuca rubra*) cv. Cascade, and Kentucky bluegrass (*Poa pratensis*) cv. Newport. *Journal of Applied Seed Production*, **1**, 39–42.

Arnold, M. H., P. C. Longdon, S. J., Brown, G. J. Curtis, R. Fletcher and **K. W. Lynch** (1984) Environment, seed quality and yield in sugar beet. *Journal of the National Institute of Agricultural Botany*, **16** 543–53.

Belteky, B. and **I. Kovecs** (1984) *Lupin, the New Break Corp*. J. G. Edwards (ed., Eng. edn). Panagri, Bradford on Avon.

Bombin-Bombin, L. M. (1980) *Seed Legislation*. Legislative Study No. 18. Food and Agriculture Organisation of the UN (FAO), Rome.

Boonman, J. G., (1971a) General introduction and analysis of problems, *Netherlands Journal of Agricultural Sciences*, **19**, 23–6. (Experimental Studies on Seed Production of Tropical Grasses in Kenya.)

Boonman, J. G. (1971b) Tillering and heading in seed crops of eight grasses, *Netherlands Journal of Agricultural Sciences*, **19**, 237–49. (Experimental Studies on Seed Production of Tropical Grasses in Kenya.)

Boonman, J. G. (1972a) The effect of row width on seed crops of *Setaria sphacelata* cv. Nande, *Netherlands Journal of Agricultural Sciences*, **20**, 22–4. (Experimental Studies on Seed Production of Tropical Grasses in Kenya.)

Boonman, J. G. (1972b) The effect of nitrogen and planting density on *Chloris gayana* cv. Mbarra, *Netherlands Journal of Agricultural Sciences*, **20**, 218–24. (Experimental Studies on Seed Production of Tropical Grasses in Kenya.)

Boonman, J. G. (1972c) The effect of time of nitrogen top-dressing on seed crops of *Setaria sphacelata* cv. Nandi, *Netherlands Journal of Agricultural Sciences*, **20**, 225–31. (Experimental Studies on Seed Production of Tropical Grasses in Kenya.)

Boonman, J. G. (1973) The effect of harvest date on seed yields in varieties of *Setaria sphacelata, Chloris gayana* and *Panicum coloratum*, *Netherlands Journal of Agricultural Sciences*, **21**, 3–11. (Experimental Studies on Seed Production of Tropical Grasses in Kenya.)

Bowring, J. D. C. and **M. J. Day** (1977) Variety maintenance for swedes and kale. *Journal of the National Institute of Agricultural Botany*, **14**, 312–20.

Brogden, A. (1977) *Tropical Pasture and Fodder Plants*. Longman, London.

Buchner, R. C. and **L. P. Bush** (eds) (1979) *Tall Fescue*. Agronomy Monograph No 20. American Society of Agronomy, Madison, Wisconsin.

Central Office of Information (1961) *Seed Improvement in Britain*. HMSO, London.

Chopra, K. R. (1982) *Technical Guidelines for Sorghum and Millet Seed Production*. FAO, Rome.

Cobley, L. S. (1977) An Introduction to the Botany of Tropical Crops, 3rd edn, rev. by W. M. Steele. Longman, London.

Cochran, W. G. and **M. C. Cox**, (1955) *Experimental Designs*. John Wiley, New York.

Concise Oxford Dictionary (1974) 5th edn, 1964, reprinted 1974. University Press, Oxford.

Copeland, L. O. (1976) *Principles of Seed Science and Technology*. Burgess, Minneapolis.

Cowling, D. W., A. F. Kelly and **Y. Demarly** (1960) *Grass Clover and Lucerne Trials*. Documentation 23, European Productivity Agency, Organisation for European Economic Co-operation, Paris.

Crane, Eva and **Penelope Walker** (1984) *Pollination Directory for World Crops*. International Bee Research Association, London.

Culfin, Claude (1986) *Farm machinery*. 11th edn. William Collins, London.

Dark, S. O. S. (1971) Cross pollination of sugar beet. *Journal of the National Institute of Agricultural Botany*, **12**, 242–66.

Delhove, G. E. and **W. L. Philpott** (eds) (1983) *World List of Seed Processing Equipment*. FAO, Rome.

Doggett, H. (1965) Cultivated sorghums. In *Essays on Plant Evolution*, Sir Joseph Hutchinson (ed.). Cambridge University Press.

Douglas, J. E. (1980) *Successful Seed Programs*. Westview Press, Boulder, Colorado.

Dyke, G. V. (1974) *Comparative Experiments with Field Crops*. Butterworth, London.

Ellerton, S. and **M. H. Arnold** (1982) Sugar beet breeding. In *50 Years of Sugar Beet Research*. International Institute for Sugar Beet Research, Brussels.

European Economic Community Council (1974) *Directive on the marketing of seed of oil and fibre plants, 1969* Official Journal of the European Communities, Brussels.

Evans, L. T. (1969) *The Induction of Flowering*. Macmillan of Australia.

Feistritzer, W. P. (ed.) (1975) *Cereal Seed Technology*. FAO, Rome.

Feistritzer, W. P. (ed.) (1982) *Technical Guidelines for Maize Seed Technology*. FAO, Rome.

Feistritzer, W. P. and **A. F. Kelly** (eds) (1978) *Improved Seed Production*. FAO, Rome.

Gane, A. J., A. J. Biddle, C. M. Knott and **D. J. Eagle** (1984), *Pea Growing Handbook*, 5th edn. Pea Growers Research Organisation, Thornhaugh, Peterborough.

George, R. A. T. (1985) *Vegetable Seed Production*. Longman, London.

Godley, C. G. A. (1975) Marketing. In *The Principles and Practice of Management*, E. F. L. Brech (ed.) Longman, London.

Gowada, C. L. L. and **A. K. Kaul** (1982) *Pulses in Bangladesh*. Bangladesh Agricultural Research Institute, Dhaka and FAO, Rome.

Gregg, Dr Bill (1983) *Seed Conditioning, Storage and Marketing*. Seed Division, Department of Agricultural Extension, Thailand.

Greenhalgh, J. F. D., I. H. McNoughton and **R. F. Thow** (eds) (1977) *Brassica Fodder Crops*. Scottish Agricultural Development Council and Scottish Plant Breeding Station, Edinburgh.

Grist, D. H. (1986) *Rice*, 6th edn. Longman, London.

Hanson, C. H. (ed.) (1975) *Alfalfa Science and Technology*. American Society of Agronomoy, Madison, Wisconsin.

Harlan, J. R. and **J. M. J. de Wet** (1972) A simplified classification of cultivated sorghum, *Crop Science*, **12**, 172–6.

Harrington, J. F. and **J. E. Douglas** (1970) *Seed Storage and Packaging Applications for India*. National Seeds Corporation and Rockerfeller Foundation, New Delhi.

Hebblethwaite, P. D. (ed.) (1980) *Seed Production*. Butterworth, London.

Hebblethwaite, P. D. (ed.) (1983) *The Faba Bean* (*Vicia faba* L.). Butterworth, London.

Hewett, P. D. (1981) Seed standards for disease in certification, *Journal of the National Institute of Agricultural Botany*, **15**, 373–84.

Hitchcock, A. A. (1935) *Manual of the Grasses of the United States*. US Government Printing Office, Washington, DC.

Hinson, K and **E. E. Harting** (1982) *Soya Bean Production in the tropics*, rev. by H. C. Minor. FAO, Rome.

House, L. R. (1985) *A guide to Sorghum Breeding*, 2nd edn. International Crops Research Institute for the Semi-arid Tropics, Patancheru AP 502 324 India.

Humphreys, L. R. (1979) *Tropical Pasture Seed Production*, 2nd printing. Plant Production and Protection Paper No. 8. FAO, Rome.

IBPGR (International Board for Plant Genetic Resources) (1980) *Descriptors for Mung Bean*. FAO, Rome

IBPGR (1981) *Descriptors for Pearl Millet, Pigeon Pea, Lupins*. FAO, Rome.

IBPGR (1982) *Descriptors for Lima Bean, Phaseolus vulgaris, Winged Bean*. FAO, Rome.

IBPGR (1983) *Descriptors for Cow Pea, Phaseolus coccineus, Safflower*. FAO, Rome.

IBPGR (1984) *Descriptors for Sorghum, Soya Bean*. FAO, Rome.

IBPGR (1985) *Descriptors for Finger Millet, Chick Pea, Faba Bean, Lentil, Vigna aconitifolia and V. tribolata, Sunflower*. FAO, Rome.

ICARDA (International Institute for Agricultural Research in Dry Areas) (1982) *Proceedings of the Seed Production Symposium, March 1981*, J. P. Srivastra and J. V. Merlin (eds). ICARDA, Aleppo, Syria.

International Commission for the Nomenclature of Cultivated Plants (1980) *International Code of Nomenclature for Cultivated Plants*. International Bureau for Plant Taxonomy and Nomenclature, Utrecht.

ISTA (International Seed Testing Association) (1971) Numbers on Seed Legislation and testing of tropical and sub-tropical seeds, *Proceedings of the International Seed Testing Association*, **36** (1).

ISTA (1984) *List of Stabilised Plant Names*. ISTA Zurich.

ISTA (1985) International rules for seed testing, *Seed Science and Technology*, **13** (2).

Jeffs, K. A. (ed.) (1986) *Seed Treatment*. CIPAC Monograph 2. Heffers, Cambridge.

Jones, D. G. and **D. R. Davies** (eds) Temperate legumes: physiology, genetics and nodulation. Pitman Books Ltd., London.

Jugenheimer, R. W. (1976) *Corn: Improvement, Seed Production and Use*. John Wiley, New York.

Kelly, A. F. and **M. M. Boyd** (1966) The stability of cultivars of grasses and clovers when grown for seed in differing environments. *Proceedings of the Xth International Grassland Congress, Helsinki*.

Kepner, R. A., R. Bainer and **E. L. Barger** (1980) *Principles of Farm Machinery*, 3rd edn. Avi Publishing Co. Westport, Conn.

Kimber, D. S. (1985) Bolting in sugar beet, *British Sugar Beet Review*, **53** (3) 15–16.

Lancashire, J. A. (ed.) (1980) *Herbage Seed Production*. Grassland Research and Practice Series No. 1. New Zealand Grassland Association, Palmerston North.

Leonard, W. H. and **J. H. Martin** (1963) *Cereal Crops*. The Macmillan Press, London.

Loch, D. S. and **G. L. Harvey** (1983) Preliminary investigations of adhesive sprays to improve seed retention in tropical grasses. *Journal of Applied Seed Production*, **1** 26–9.

Longdon, P. C. (1972) Monogerm sugar beet seed production experiments, *Journal of Agricultural Science*, **78**, 497–503.

Van Marrewijk and **H. Toxopeus** (eds) (1979) *Eucarpia Cruciferae 1979 Conference*. Foundation for Agricultural Plant Breeding and Government Institute for Research on Varieties, Wageningen.

Martin, Hubert (1973) *The scientific principles of crop protection*, 6th edn. Edward Arnold, London.

Mayo, O. (1980) *The Theory of Plant breeding*. Clarendon Press, Oxford.

McLean, K. A. (1980) *Drying and Storing Combinable Crops*. Farming Press, Ipswich.

Molina, E. C. (1949) *Poisson's Exponential Binomial Limit*. D. van Nostrand, New York.

Neargaard, Paul (1977) *Seed Pathology*, vols 1 and 2. The Macmillan Press, London.

NIAB (National Institute of Agricultural Botany) (1975) *Growing Kale for Seed*. Seed Growers Leaflet No. 2. NIAB, Cambridge.

NIAB (1979) *Growing Turnip for Seed*. Seed Growers Leaflet No. 3. NIAB, Cambridge.

NIAB (1980) *Growing Cabbage for Seed*. Seed Growers Leaflet No. 4. NIAB, Cambridge.

NIAB (1980) *Growing Grasses and Herbage Legumes for Seed*. Seed Growers Leaflet No. 5. NIAB, Cambridge.

NIAB (1981) *Growing Faba Beans for Seed*. Seed Growers Leaflet No. 6. NIAB, Cambridge.

NIAB (1982) *Growing Peas for Seed*. Seed Growers Leaflet No. 7. NIAB, Cambridge.

NIAB (1982) *Growing Italian Ryegrass and Tetraploid Hybrid Ryegrass to Obtain Optimum Seed Yields and Quality*. Seed Growers Leaflet No. 8. NIAB, Cambridge.

OECD (Organisation for Economic Co-operation and Development) (1973) *Methods for Plot Tests and Field Inspection*. OECD, Paris.

OECD (1977) Herbage and oil seed scheme; cereal seed scheme; maize seed scheme; sugar beet and fodder beet seed scheme; scheme for seed of subterranean clover and similar species. In *OECD Schemes for the Varietal Certification of Seed Moving in International Trade*. OECD, Paris.

OECD (1982) *Methods for Plot Tests and Field Inspection*. OECD, Paris.

OECD (1984) *List of Cultivars Eligible for Certification*. OECD, Paris.

Percival, Mary (1965) *Floral Biology*. Pergamon Press, London.

Puri, S., C. D. Ennis and **K. Mullen** (1979) *Statistical Quality Control for Food and Agricultural Scientists*. G. K. Hall, Boston.

Purseglove, J. W. (1984) *Tropical Crops: Dicotyledons*, 3rd impression, reprinted 1984. Longman, London.

Purseglove, J. W.(1985) *Tropical Crops: Monocotyledons*, 5th edn. Longman, London.

Rosell, C. H. and **A. F. Kelly** (eds) (1983) *Seed Campaigns*. FAO, Rome.

Rotzoll, H. (1982) Seed marketing. In *Seeds*, W. P. Feistritzer (ed.). FAO, Rome.

Shaw, N. H. and **W. W. Bryan** (eds) (1976) *Tropical Agricultural Research*. Commonwealth Agricultural Bureaux, Farnham Royal.

Simmonds, N. W. (1981) *Principles of Crop Improvement*. Longman, London.

Simmonds, N. W. (ed) (1984) *Evolution of Crop Plants*. Longman, London.

Sinka, S. K. (1977) *Food Legumes: Distribution, Adaptability and Biology of Yield*. Plant Production and Protection Paper No 3. FAO, Rome.

Skerman, A. V. (1977) *Tropical Pasture and Fodder Plants*. FAO, Rome.

Sneddon, J. L. (1963) Sugar beet seed production experiments. *Journal of the National Institute of Agricultural Botany*, **9** 333–45.

Sneddon, J. L. (1964) Seed Production experiments with turnip. *Journal of the National Institute of Agricultural Botany*, **10** 116–21.

Steel, H. E. and **H. Opilu** (1976) *The Physiology of Flowering Plants*. Edward Arnold, London.

Summerfield, R. J. and **E. H. Roberts** (eds) (1985) *Grain Legume Crops*. Collins, London.

Summerfield, R. J. and **A. H. Bunting** (eds) (1980) *Advances in Legume Science*. Royal Botanic Gardens, Kew.

Teen, C. C. (1974) *Triticale, First Man-made Cereal*. American Association of Cereal Chemists, St Paul, Minn.

Thomson, J. R. (1979) *An Introduction to Seed Science and Technology*. Leonard Hill, London.

United States Department of Agriculture (1961) *Seeds; the Year Book of Agriculture*. USDA, Washington, DC.

UPOV (International Union for the Protection of New Varieties of Plants) (1973) Lucerne (*Medicago sativa* L. and *M.* × var. Martyn); red clover (*Trifolium pratense* L.); runner beans (*Phaseolus coccineus* L.) In Guidelines for the Conduct of Tests for Distinctness, Homogeneity and Stability. UPOV, Geneva.

UPOV (1974) Wheat (*Triticum durum* Desf.). In *Guidelines for the Conduct of Tests for Distinctness, Homogeneity and Stability*. UPOV, Geneva.

UPOV (1976) Bent (*Agrostis canina* L., *A. gigantea* Roth, *A. stolonifera* L. and *A. tenuis* Sibth); common vetch (*Vicia sativa* L.); turnip (*Brassica rapa* L. var. *rapa*). In *Guidelines for the Conduct of Tests for Distinctness, Homogeneity and Stability*. UPOV, Geneva.

UPOV (1979a) Lupins (*Lupinus albus L., L. augustifolius* L. and *L. lutens* L.). In *Guidelines for the Conduct of Tests for Distinctness, Homogeneity and Stability*. UPOV, Geneva.

UPOV (1978a) *International Convention for the Protection of New Varieties of Plants*. UPOV, Geneva.

UPOV (1978b) Rye (*Secale cereale* L.). In *Guidelines for the Conduct of Tests for Distinctness, Homogeneity and Stability*. UPOV, Geneva.

UPOV (1979a) Lupins (*Lupinus albus, L. augustifolius* L. and *L. lutens* L.). In *Guidelines for the Conduct of Tests for Distinctness, Homogeneity and Stability*. UPOV, Geneva.

UPOV (1979b) *Revised General Introduction to the Guidelines for the Conduct of Tests for Distinctness, Homogeneity and Stability*. UPOV, Geneva.

UPOV (1980) Flax, linseed (*Linum usitatissimium* L.); maize (*Zea mays* L.); ryegrass (*Lolium multiflorum* Lam, *L. perenne* L. and hybrids); sheep's fescue (including hard fescue) and red fescue (*Festuca ovina* L. *sensu lato* and *F. rubra* L.). In *Guidelines for the Conduct of Tests for Distinctness, Homogeneity and Stability*. UPOV, Geneva.

UPOV (1981) Barley (*Hordeum vulgare* L. *sensu lato*); oats (*Avena sativa* L. and *A. nuda* L.); peas (*Pisum sativum* L. *sensu lato*); wheat (*Triticum aestivum* L.), In *Guidelines for the Conduct of Tests for Distinctness, Homogeneity and Stability*. UPOV, Geneva.

UPOV (1982) French bean (*Phaseolus vulgaris* L.). In *Guidelines for the Conduct of Tests for Distinctness, Homogeneity and Stability*. UPOV, Geneva.

UPOV (1983) Soya bean (*Glycine max* (L.) Merrill); sunflower (*Helianthus annuis* L. and *H. debilis* Nut). In *Guidelines for the Conduct of Tests for Distinctness, Homogeneity and Stability*. UPOV, Geneva.

UPOV (1984) Broad bean, field bean (*Vicia faba* L.); cocksfoot (*Dactylis glomerata* L.); meadow fescue and tall fescue (*Festuca pratensis* Huds and *F. arundinacea* Schreb); rice (*Oryza sativa* L.); swede (*Brassica napus* L. var. *napobrassica* Rehb); timothy (*Phleum pratense* L. and *Ph. bertolonii* D. C.). In *Guidelines for the Conduct of Tests for Distinctness, Homogeneity and Stability*. UPOV, Geneva.

UPOV (1985) Groundnut (*Arachis* L.); white clover (*Trifolium repens* L.). In *Guidelines for the Conduct of Tests for Distinctness, Homogeneity and Stability*. UPOV, Geneva.

Vines, A. E. and **N. Rees** (1972) *Plant and Animal Biology*, vol. 2. Pitman Publishing, London.

Ward, J. T., W. D. Basford, J. H. Hawkins and **H. M. Holliday** (1985) *Oilseed Rape*. Farming Press, Ipswich.

Weiss, E. A. (1983) *Oilseed Crops*. Longman, London.

Wheeler, W. A. and **D. D. Hill** (1957) *Grassland Seeds*. D. van Nostrand, New York.

WPBS (Welsh Plant Breeding Station) (1978) *Principles of Herbage Seed Production*, 2nd edn, rev. E. W. Bean. WPBS, Aberystwyth.

Zaleski, A. (1970) White clover seed production. In *White Clover Research*, J. Lowe, (ed.). Occasional Symposium No. 6, British Grassland Society, Reading.

Zohary, M. and **D. Heller** (1984) *The Genus Trifolium*. The Israel Academy of Sciences and Humanities, Jerusalem.

INDEX